Ramsay Wright

An introduction to zoology for the use of High schools

Ramsay Wright

An introduction to zoology for the use of High schools

ISBN/EAN: 9783337124670

Printed in Europe, USA, Canada, Australia, Japan

Cover: Foto ©berggeist007 / pixelio.de

More available books at **www.hansebooks.com**

AN INTRODUCTION

TO

ZOOLOGY.

FOR THE USE OF HIGH SCHOOLS.

BY

R. RAMSAY WRIGHT, M.A., B. Sc.

Professor of Biology in the University of Toronto.

Toronto:

THE COPP, CLARK COMPANY (LIMITED).

1889.

PREFACE.

The present volume has been prepared with the object of aiding the study of Zoology in the Ontario High Schools. Already, one branch of Natural History—Botany—has been introduced with gratifying results, and it is thought that the addition of the elements of Zoology to the course may similarly awaken a wide-spread interest in animal life throughout the Province.

The plan of treatment adopted is substantially that of the Syllabus prescribed by the Education Department. Attention is first directed to the Vertebrates as the most familiar and conspicuous animals, but the essential characteristics of the chief groups of Invertebrates are also given, the greater amount of space being, however, devoted to such groups as have terrestrial or fresh-water representatives. In each of the classes of the Animal Kingdom, some easily obtainable form is employed as a type in which to point out the more obvious structural features of the class, and it is assumed that these will be verified by actual examination.

A number of figures have been introduced, partly with the view of facilitating such examination, partly to illustrate the less accessible forms. These have, for the most part, been copied from scientific works like the publications of the U. S. Fish Commission, Brehm's *Thierleben*, etc., but a few have been drawn for the occasion, Figs. 1 and 58 with several others, being from the pen of Mr. E. E. Thompson.

Much of the educational value of Botany as generally taught in schools results from the accurate observation necessary to employ the terminology correctly, and to make correct diagnoses: Zoology does not lend itself so easily to this kind of exercise, but it affords an equally valuable discipline—the tracing of the modifications of form throughout less nearly allied groups. A good deal of space has, accordingly, been devoted to this aspect of Zoology, although other aspects which may excite the interest of the young student have not been neglected. For example, the chapters on the Reptiles and Birds give prominence to the remarkable geological history of these classes ; that on the Mammals, to the correlation of form and habit in the group ; while the last chapter aims at showing the connection of the various subdivisions of zoological study.

Experience alone will show, what form zoological instruction in the Secondary Schools ought to assume, so as not to interfere with other departments of study : the text-books on the "type-system" seemed to be too advanced for the present purpose, and also not to afford as wide an acquaintance with the forms of Animal Life as is desirable, while many elementary, systematic text-books prepared for school use do not demand the actual examination of types, so necessary for the formation of clear conceptions. It is hoped that the present volume which endeavors to combine the advantages of both systems may prove adapted to the purpose for which it is intended.

Toronto. July. 1889.

CONTENTS.

HIGH SCHOOL ZOOLOGY.

CHAPTER I.

THE STRUCTURE OF THE CATFISH.

1. Botanical students will remember that plants are often subdivided into **phanerogamic** and **cryptogamic** forms ; the latter lack a certain characteristic way of producing seeds which is present in the former, but they really embrace several distinct primary subdivisions of the Vegetable Kingdom, whose only common character is the negative one referred to above. Similarly the Animal Kingdom is often subdivided into **Vertebrate** and **Invertebrate** animals, but the latter really include several distinct sub-kingdoms sharing the negative character of the absence of a backbone. Although, then, the Botanist and Zoologist regard the terms cryptogamic and invertebrate as survivals from a period when less was known as to the structure of the contained forms than there is now, yet the terms are very convenient for every-day use, because they separate the less important, *i. e.*, the lower and less conspicuous members of the Vegetable and Animal Kingdoms f.om those which are not only higher and more familiar, but also more economically important.

2. The history of Botany and Zoology teaches us that for various reasons these sciences have progressed most rapidly at first with the study of the higher forms of life : similar reasons will render it more convenient for us to begin our study of Zoology with an examination into the structure of a Vertebrate

2

Animal. In making our selection, however, it will be desirable to choose a form which shall be so far typical, that a knowledge of the structure of its various organs will enable us to interpret the nature and significance of the comparable or **homologous** parts in other Vertebrates. No one animal is best in every respect for this purpose, because there is no animal which unites in itself all the characters which we regard as **primitive** or **general.** An example will render the meaning of these terms plain. Most Vertebrates have five fingers on the hand, and we regard that as a primitive or general arrangement in comparison with that in a cow where there are two, or in a horse where there is only one. Such a reduction in number we regard as a **specialization** associated with the function which the hand performs, and it is very much easier to interpret correctly the specialized condition if we have in the first place familiarized ourselves with the more primitive one. Our object must then be to find some fairly primitive form, which is common, easily obtained, and easily studied : our demands in all these respects are pretty well met by the common catfish, the angling for which is attended by no great difficulties, which is tenacious of life and easily kept in captivity, and which finally occupies such a place in the class of the Fishes that we can, after acquainting ourselves with its structure, survey the other members of the class, and proceed to the study of the higher Vertebrates.

3. **General Form.**—All Vertebrates, like most Invertebrates, are **bilaterally symmetrical** animals, i. e., the body is divisible into right and left symmetrical halves by a plane passing from head to tail through the middle line of the back (or **dorsal** surface) as well as through the middle of the lower (or **ventral**) surface. This is the median **sagittal** plane ; planes at right angles to it, which are parallel to the dorsal and ventral surfaces, are called **horizontal,** while those which **transect** the body at right angles to both are **frontal.**

Different regions of the body have different duties to discharge, and consequently differ in form and structure. We distinguish in a fish the head, trunk, and tail, of which the first lodges the brain and sense organs, secures food, and shelters the gills, the last is chiefly locomotive in function, while the trunk differs from both in being hollowed out so as to enclose the intestines and other viscera in the so-called body-cavity (cœlom). The regions referred to are said to be **axial,** because they are disposed round the chief axis of the body, while the two pairs of limbs or appendages, much more developed in the higher Vertebrates, project laterally from the trunk, to which they are attached in the neighbourhood of the cephalic and caudal regions respectively, and are described as **appendicular.** In a fish the anterior and posterior appendages are known as the pectoral and ventral fins, (Fig. 1) from which are to be

Fig. 1.—Common Catfish, or Bullhead. ⅓.
Amiurus nebulosus.

distinguished the unpaired fins, occupying the middle line of the dorsal and ventral aspects of the trunk and tail, and assisting in locomotion. The latter are named from their position dorsal, caudal and anal. In the catfish, part of the dorsal is separated as the **adipose** fin, which is regarded as the rudiment of a longer dorsal, and, instead of being supported by fin-rays, has only fatty tissue within it.

4. Apertures.—Certain apertures exist on the surface of the body; of these, the mouth is bounded by the upper and lower jaws and leads into the mouth-cavity, the nostrils or openings of the olfactory sacs are four in number on the dorsal surface of the head, but there are no external apertures for the ears. On the ventral surface in front of the anal fin is the posterior aperture of the intestine, and immediately behind it that of the urinary and reproductive organs. On the sides of the head behind the mouth are certain large apertures—the **gill-slits**, five in number, opening into the cavity of the mouth, but these are ordinarily concealed by the gill-cover or **operculum**, a flap which projects backward over them, and by a membrane attached to the inner surface of the flap, the **branchiostegal** membrane. In this way the gill or **branchial** chamber is formed, opening by the **branchial aperture** along the hinder and lower border of the said flap. Both the branchiostegal membrane and the gill-cover have a supporting framework of bones, the branchiostegal rays in the one case, the opercular bones in the other.

In addition to the olfactory organs referred to above, the following sense organs are to be noted : the eyes ; certain holes and slits along the lateral line and on the head leading into canals and pits in which sense organs are situated ; and the **barbels**, sensitive processes of the skin of the head, eight in number in the catfish, but frequently absent in other forms.

5. In most fishes the skin is strengthened by bony scales, either round in outline—**cycloid**—or with the hinder margin toothed—**ctenoid**—(Fig. 2), but the catfish is destitute of such, except for certain very minute ones which are in the walls of the lateral canal. The skin is therefore soft and slippery, and variously coloured according to the distribution of pigment in it. It is tightly bound down to the underlying flesh by slips of fibrous tissue, but in certain parts some loose **subcutaneous** tissue is accumulated between them. When a sharp cut is made through the skin it is possible to recognize two layers, an outer, the **epidermis** and an inner, the **corium**.

which layers indeed exist in the skin of all Vertebrates. It is chiefly but not exclusively in the latter that the pigment is contained. If the epidermis be removed by scraping, the exposed surface of the corium will be observed to be rough with

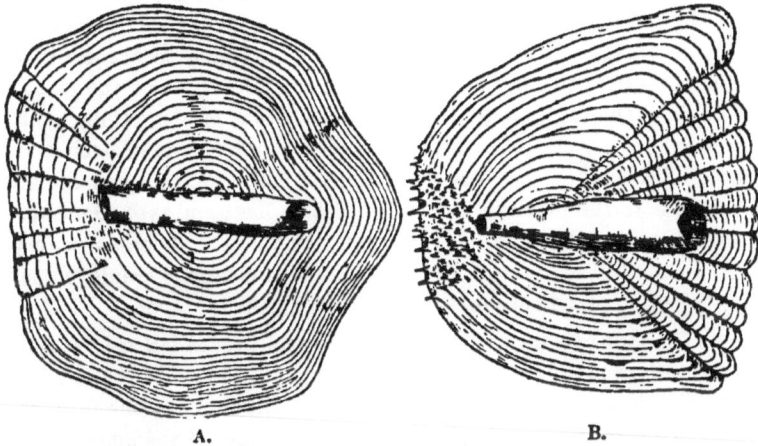

Fig. 2.—A, Cycloid Scale from Lake Herring.   B, Ctenoid Scale from Rock Bass. $^6/_1$.

papillæ, so that the smoothness of the surface is due to these interpapillary spaces being filled up with the epidermis. The papillæ are of importance as the channels through which nerves and nutriment from the blood-vessels reach the epidermis from the corium.

6. Minute Structure of the Skin.

GENERAL REMARKS ON HISTOLOGY.—That branch of the study of the structure of animals which deals with the minute elements of the various organs, and which requires the microscope and other tools in the course of its investigations, is termed **Histology.** Each organ of the body is built up of **tissues,** and each tissue is formed of ultimate elements named **cells** arranged in a characteristic way. Thus the skin, which is a complicated organ performing very different functions, is composed chiefly of two kinds of tissues—the **epithelial** tissue of the epidermis, and the **connective** tissue of the corium. The former may fairly be considered its most characteristic tissue, but both are necessary elements in its structure. It is still more complex, however, in virtue of the presence of both **muscle** and **nerve** tissue, so that this one organ of the body contains tissues of all the four categories under which histologists arrange the component parts of the animal body.

7. Animal cells have the same component parts as vegetable cells, *i. e.*, they are formed of a protoplasmic cell-body containing a nucleus and limited (frequently) by a membrane or wall. The latter, which plays so important a part in the support of the plant, resigns this function in the higher animals to the intercellular substance, which, although like the cell-wall formed by the activity of the protoplasm, differs therefrom in rarely exhibiting the territories belonging to the constituent cells. All the cells of the animal body are, like those of the plant, derived from the division (and the differentiation of the products of the division) of one cell—the egg-cell, and the first results of such division and differentiation are the formation of embryonic layers somewhat analogous to the primary meristems of the plant-embryo. Perhaps the most characteristic difference between the plant and animal embryo is that in the latter some of the most important organs are developed by the infolding of the originally superficial epithelial layer.

8. The four categories under which animal tissues fall may shortly be characterized as follows :—(Fig. 3.)

I. **Epithelial Tissue** is that which is disposed in the form of one or more layers of distinct cells on the free surfaces of the body, including the alimentary canal, the lining of the cœlom, the cavities of the nervous system, etc. The cells may be cylindrical, columnar, cubical or scale-like in form, their free surfaces may be covered with a resistant cuticle, or provided with delicate continuations of the protoplasm in the form of cilia or hairs. If their duty is to receive impressions and transmit them to nerves, they constitute neuro-epithelium ; if they secrete some characteristic product they constitute glandular epithelium, and are generally turned in from the free surface for protection ; if they serve merely to form hard structures for protection of underlying parts or for defence, they are modified into horny epithelial scales, feathers, hairs, hoofs, nails, horns, etc., while if they are converted into eggs, etc., they constitute germinal epithelium.

II. **Connective Tissues.**—These constitute the framework of the body, which in some organs is of the utmost delicacy, in others, the true skeletal tissues, attains great firmness and hardness. Sometimes the cellular elements are distinct, in which case they may be free to wander through the interspaces of the tissues in the form of amœboid or wandering cells, or be more limited in their mobility like the pigment-cells; or be fixed and flat like epithelium, or globular and filled with fat, or branched and communicating with their neighbours. Sometimes in the adult tissue the protoplasm is almost all converted into intercellular

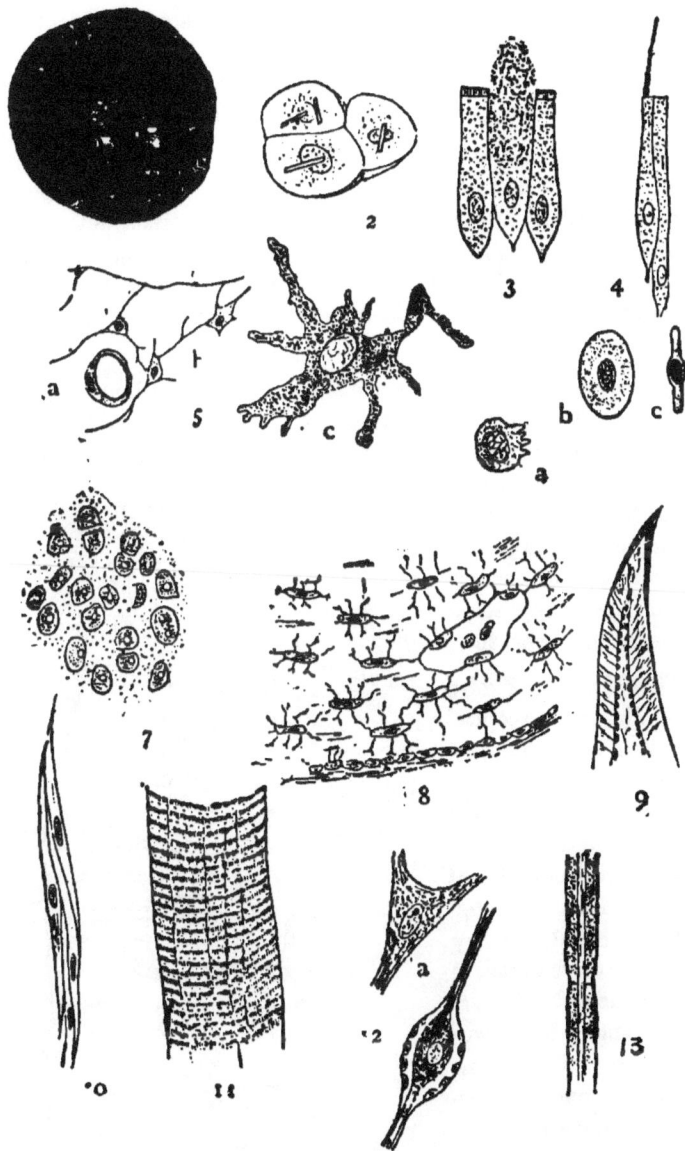

Fig. 3.—Illustrations of the Simple Tissues of the Catfish

1. Egg, as type of an animal cell.   2. Scale-like epithelial cells from skin.   3. Columnar ones from intestines.   4. Neuro-epithelial and supporting cells from ear.   5. Loose connective tissue formed of branched cells (b) with fat (a) and pigment-cells (c). 6. Blood cells ; a, colourless, b and c coloured.   7. Cartilage.   8. Bone.   9. Tooth with pulp-cavity, dentine and enamel-cap.   10. Simple muscle-cells.   11. Striated muscle-fibre.   12. Nerve-cells.   13. Nerve fibre.

substance, in which case nuclei with a scant surrounding film of proto-plasm are left embedded in a matrix which may be soft and jelly-like, or converted into fibres, or stiff and homogeneous like cartilage, or hard through incorporation with lime-salts like bone and tooth.

III. **Muscle-tissue.**—The cellular character is to be seen in the simpler kind which is formed of much elongated cells, whose protoplasm is highly contractile, while the higher kind of muscle-tissue is that termed **striated** (from the appearance of the contractile substance of the fibres under the microscope). In the latter, each fibre is a unit of higher rank than the simple muscle-cell, because it is the equivalent of several cells.

IV. **Nerve-tissue.**—Two elements are distinguished, nerve-cells and nerve-fibres. The cellular character of the former is always evident; the latter are to be regarded as processes of these cells, each nerve-fibre having for its core an **axis-cylinder** continuous with the cell protoplasm, which serves for conduction, and is generally isolated by one or more sheaths.

9. Histology of the Skin. — The microscopical examination of a thin pre-pared section of skin dis-closes at once the two chief component parts (Fig. 4). Of these the horizontal fibres of the corium are separated from the epithelial layers by a looser connective tissue, in which pigment cells are abundant and which pro-jects into the papillæ. Some looser fatty connective tis-sue may separate the hori-zontal fibre-layers from the flesh. Filling up the inter-spaces between the papillæ are the epithelial cells of

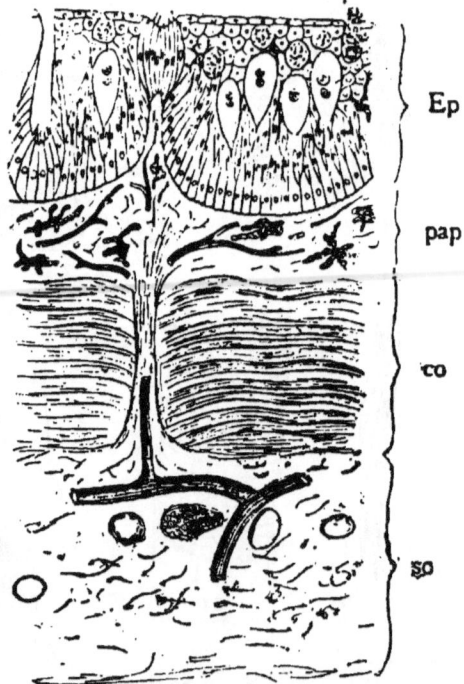

Fig. 4.—Diagram of section of the Skin in the Cat-fish. Ep, epithelium; pap, papillary layer of the corium, co; sc, subcutaneous connective tissue, with nerves and blood-vessels. ×50.

the epidermis ; these present several varieties in shape and function, those next the corium being columnar, those on the free surface cubical, while the intervening cells are intermediate in shape. Certain peculiar cells stand out from the others ; these are the slime-cells present in all fishes, which provide the skin with its covering of mucus, and the clavate cells of unknown function, occurring chiefly in fishes with a soft skin like the Eel and Burbot. A few pigment-cells wander out from the corium into the interspaces of the epidermis. Finally certain special epithelial cells are to be found in bud-like groups on the end of some of the papillæ. Each of these is provided with a delicate hair-like process at its tip, and is connected with a nerve at its basal end ; they thus belong to the class of neuro-epithelial cells and they constitute the simplest form of sense-organ in the fish. It is supposed that they are affected by vibrations in the surrounding medium, and that they are tactile in function. That this is so may be inferred from the use to which the barbels or feelers, which are covered with these organs, are put. Elsewhere on the surface of the fish, groups of similar cells occur, not projecting freely on the surface, but retracted for protection into minute sacs opening by slits on the surface, or projecting at intervals into the cavities of the sensory canals of the lateral line and head, which communicate with the surrounding medium by distinct pores. The button-like hillocks of neuro-epithelium are generally protected by a bony scale, and it is the fusion of such scales which gives rise to some of the skin-bones of the head.

10. **Skeletal System.**—In the course of the above paragraphs reference has been made to bones developed and situated in the skin. When such bones acquire no connection with the deeper parts they are said to belong to the **exoskeleton ;** the rough teeth in shark's skin, and the scales of a white-fish, *e. g.*, are of this nature. It is evident from what has been said that the catfish is very poorly provided with exoskeletal structures ; all of its bones belong to the internal or **endoskeleton.**

11. The skeletal system is formed chiefly by tissues of the connective-tissue group, viz., fibrous connective tissue (in the form of ligaments and membranes), cartilage and bone. Of these the cartilage plays only a transitory part in the development of the catfish's skeleton, both it and the fibrous connective tissue being in great part converted into bone. Teeth are true exoskeletal structures, although they are only found in

connection with the internal skeleton in the catfish : we know this by the development of their two constituent parts, the dentine—a member of the connective-tissue group allied to bone—and the enamel, which is formed by the modification of epidermal cells. (Fig. 3.)

12. In the skeleton we distinguish axial and appendicular parts, § 3 ; to the former belong the skull with the hard parts of the gills, and the vertebral column with the ribs.

We shall study the latter first ; it is formed of a series of separate bones, the **vertebræ**, which vary considerably in form in different regions of the column, but are all characterized by a central part (the body or **centrum**), which is hollowed out like a cup, on both anterior and posterior faces, (**amphicœlous.**) (Fig. 5.)

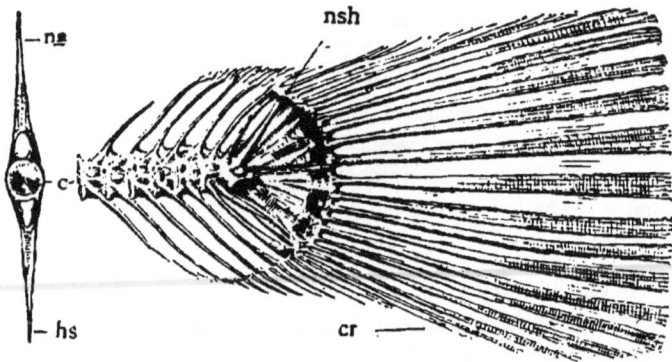

Fig. 5.—Caudal Vertebra and Caudal end of Vertebral Column in the Catfish. Ns. neural spine ; c, vertebral centre ; hs, hæmal spine ; nsh, bony sheath of the notochord ; cr, caudal rays.

Within the space so formed is contained the gelatinous remains of the **notochord,** a rod present in the youngest stages of all Vertebrates, around which the vertebral column is built, but rarely continuous in the adult, except in the lowest Vertebrates.

Each centrum bears on its dorsal surface an arch, the **neural** arch, which terminates in a neural spine. The series of centra constitutes a flexible rod of great importance in loco-motion, the series of neural arches forms a canal serving to pro-tect that part of the nervous system known as the spinal cord, while the series of spines (to which the common name of spine, spinal column, etc., for the vertebral column is due) chiefly serves for the attachment of muscles. In higher Vertebrates not only are the centra intimately united to each other, but the arches have certain projections (**articular processes**), forming joints with similar processes on the arches in front and behind. These are not very much developed in the catfish, but this mode of union permits a certain amount of rotary movement between the vertebræ, to which the word vertebra owes its origin.

In all fishes there may be distinguished in the vertebral column two regions, the trunk and the tail, the former extending as far back as the cœlom referred to above, the latter behind that. In addition to the neural arches and spines, the caudal vertebræ have **hæmal** arches and spines, which protect the blood vessels running back through the tail below the centra, while the trunk vertebræ have ribs which generally protect the contents of the cœlom, but do not meet below nor carry spines.

The ribs are not articulated directly to the centra, but to projections from them, **transverse processes**, which appear to be the real representatives of the hæmal arches.

Some of the anterior trunk vertebræ in the catfish and its allies are very different from the others, being modified in connection with the organ of hearing, and in all fishes the most posterior vertebræ are altered in connection with the tail. The special way in which this alteration affects the catfish may be seen in **Fig. 5,** as far as appearance goes the rays of the caudal fin seem to be equally divided above and below the end of the vertebral column, (the fin is said to be **homocercal**), but in reality the tip of the vertebral column turns abruptly upwards so that most of the rays are really on its ventral surface. We shall see that in certain other fishes this unequal division of the tail-fin is much more apparent (**heterocercal**).

14. It is in the vertebral column that the segmentation or **metamery** of the Vertebrate body finds its most evident expression, but we shall find that many other organs are likewise divided into **metameres**, notably the muscular and nervous systems.

15. A study of the development of the catfish shows that not only is the vertebral column built around the notochord, § 12, but also that the same is true of a considerable part of the skull, and that the notochord in the head is related in the same way to the nervous system and alimentary canal as it is in the trunk. It was at one time thought that it should be possible to distinguish constituent vertebræ in the skull, but this is impossible, for the very different functions which the anterior part of the axial skeleton has to discharge are associated with corresponding differences in form. Thus the fact that the anterior end of the central nervous system is dilated into the brain, is associated with the development of a sheltering box, the **cranium,** which is further modified by the apposition or incorporation with it of the protecting hard parts of the higher sense-organs, and the fact that the anterior end of the alimentary canal is devoted to securing food and to respiration is associated with the development of certain hard parts in connection therewith—the **visceral skeleton.** Of these we shall study first—

16. **The Cranium.**—The cranial box has certain openings, one (the occipital foramen or **foramen magnum**) to permit of the spinal cord joining the brain, others to allow the escape of nerves and the entry or egress of blood vessels. Although originally in the young fish largely cartilaginous in its texture, the box afterwards becomes partly converted into bone, and the bones formed in the cartilage are related in a definite way to these openings. Other bones are formed in the skin for the protection of the sensory canals; still others (especially in the roof of the mouth) for the support of the teeth, and both of

these kinds may be closely incorporated with those of the first category.

17. An inspection of the cranium from the upper surface (Fig. 6) discloses in the middle line two slits (the anterior and

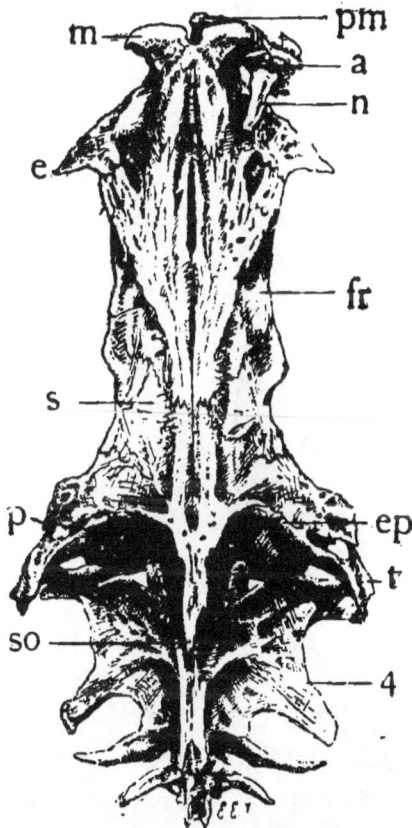

Fig. 6.—Cranium and Anterior Vertebræ of Catfish from above.

M. mesethmoid ; pm. premaxilla ; a. antorbital ; n. nasal ; e. parethmoid ; fr. frontal ; s. sphenotic ; p. pterotic ; ep. epiotic ; t. supraclavicle ; so. supraoccipital spine ; 4. transverse process of fourth vertebra.

posterior **fontanelles**) which separate the two **frontal** bones except for a short intervening bridge, where these bones (which form a considerable part of the cranial roof) articulate by a serrated suture. In front of the anterior fontanelle is the **mesethmoid** bone : behind the posterior, the **supra-occipital**, the former terminating anteriorly in a notch, the latter posteriorly in a bifid spine.   Four projections mark each lateral border of the skull, belonging to the mesethmoid, the **parethmoid,** the **sphenotic** and the **pterotic** bones, while the hinder border presents a projection from the **epiotic** on either side of the supra-occipital spine.   Thus twelve bones enter into the formation of the cranial box, for all of the above-mentioned bones are in pairs with the exception of the mesethmoid and supra-occipital.

18. On the floor of the skull **(Fig. 7)** the mesethmoid still forms

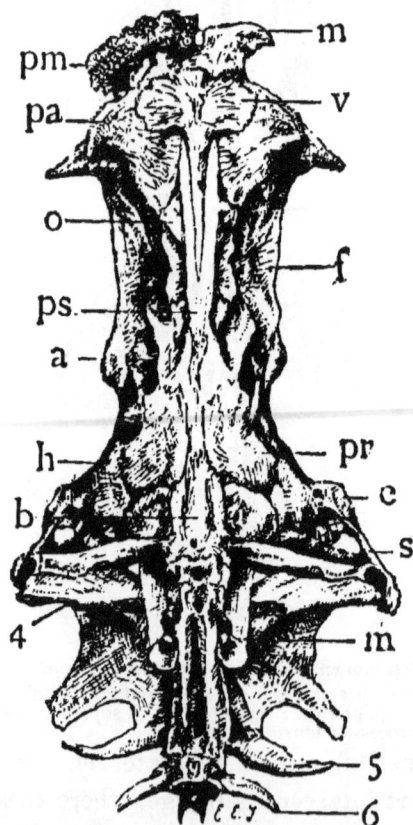

Fig. 7.—Cranium and Anterior Vertebræ of Catfish, from below.

Pm, premaxila ; m, mesethmoid ; v, vomer ; pa, parethmoid ; o, orbitosphenoid ; f, frontal; ps, parasphenoid; a, alisphenoid; pr, prootic; h, articular surface for hyomandibular on sphen- and pterotics ; b, basioccipital with exoccipitals on either side ; s, supraclavicle ; m, "malleus;" 4, 5 and 6, transverse processes of 4th, 5th and 6th vertebræ.

the anterior boundary, while the posterior is occupied by another unpaired bone, the **basi-occipital**. The latter articulates with the centrum of the first vertebra, and also affords support to a bone of the shoulder-girdle (the **supraclavicle** or **post-temporal**) which abuts against the strong transverse process of the fourth vertebra and rests by two other prongs on the epi- and pterotic projections. Between the mesethmoid and the basioccipital the following bones are to be noticed; the **vomer**, a T-shaped bone, the transverse bar of which lies across the ventral surface of the parethmoids, and the **parasphenoid** which continues back the leg of the vomer in the middle line. All of the lateral projections seen from above are also to be seen from below. Certain foramina furnish landmarks for the recognition of the other bones to be seen on this surface: on either side of the basioccipital are the **exoccipitals** each with two foramina for the escape of the 9th and 10th nerves. The sutures between the basi- and exoccipitals, and the epi- and pterotics are occupied by cartilage and are thus very distinct, and the **pro-otics** (which are wedged in between the exoccipitals and the pterotics) are similarly marked out. In front of the pro-otics are the large apertures by which the 5th and 7th cranial nerves escape, but these apertures are also partly bounded by the sphenotic above, the **alisphenoid** in front and the **basisphenoid** (an unpaired bone partly concealed by the parasphenoid) towards the middle line. Between the alisphenoids and the **orbitosphenoid,** are the optic foramina for the escape of the optic nerves ; the latter bone is unpaired, and it forms a considerable part of the side walls of the skull and also of its floor, where, however, it is covered by the parasphenoid : it is also channelled by the olfactory tracts on their way to the nasal sacs. Thus eleven additional bones are to be seen from this aspect, of which five are unpaired; viz., the basioccipital, the basi-, orbito-, and parasphenoids, and the vomer.

19. The shape of the cranial cavity will be better understood after a description of the brain and ear, but the cavity is narrower and shallower in front where the olfactory tracts are alone to be accommodated, and wider and deeper behind where the brain and ear are situated.

20. Reference was made above to some scale-like bones in connection with the sensory canals. Several of these form an infraorbital chain below the eye; two of them, the **antorbital** and **nasal**, are in the roof of each nasal sac, while others have been incorporated with the underlying frontals, parethmoids, etc.; for these and several other bones of the roof shelter sensory canals.

21. **The Jaws and Visceral Skeleton.**—Immediately below the lateral edge of the sphenotic is a groove lined with cartilage, which extends on to the pterotic, and is the articular surface for the **hyomandibular**, an important bone suspending the jaws and the visceral skeleton to the skull. (Fig. 8). Inti-

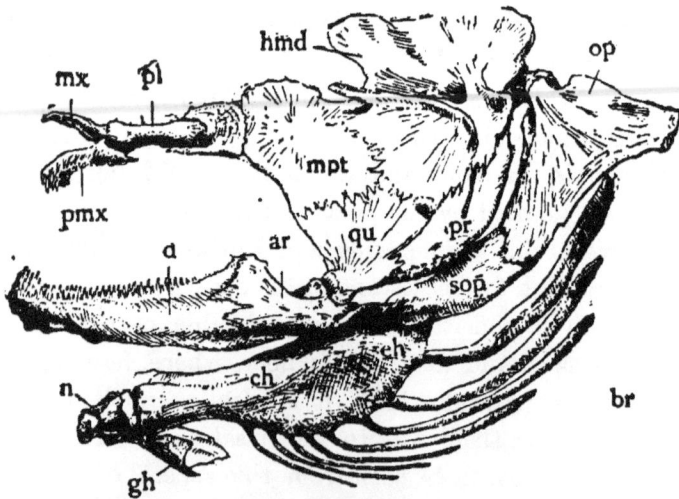

Fig. 8.—Jaws and Hyoid Arch of Catfish, from the side.

Mx, maxilla; pmx, premaxilla; pl. palatine; hmd, hyomandibular; op, operculum; mpt, metapterygoid; qu, quadrate; pr, preoperculum; sop, interoperculum; d, dentary; ar, articular; h, hypohyal; gh, glossohyal; ch, ceratohyal; eh, epihyal: br, branchiostegal rays.

mately united to it by an intervening symplectic cartilage is the **quadrate**, a bone which furnishes the articular surface for the lower jaw or **mandible**. Wedged in between the hyomandibular and quadrate is a flat bone, the **metapterygoid**, to the anterior end of which (by means of an intermediate scale-like bone,) the **palatine** is related, a rod-like bone articulated to the parethmoid and anteriorly carrying the **maxilla**. In most fishes this bone forms part of the gape ; here it acts merely as a support for the large maxillary barbels, while the **premaxillæ** attached to the ventral surface of the horns of the mesethmoid, and connected with each other in the middle line, bear most of the teeth of the upper jaw.

22. The mandibles also bear teeth on their so-called **dentary** part, while near the "angle" of the jaw is the **articular** part ; on the inner surface of both the remains of the cartilage (**Meckel's**) on which this jaw is built are to be seen.

23. Closely united to the hinder border of the hyomandibular and quadrate is the **preoperculum** through which a sensory canal runs to reach the lower jaw. Behind it is the moveable **operculum**, the chief bone of the gill-cover, which however articulates separately with the hyomandibular. Between the operculum and the mandible, and united with both, is the third bone of the gill-cover, the **interoperculum**. In most fishes, but not here, there is a fourth bone, the **suboperculum**.

24. By means of a short **interhyal** piece of cartilage the hyoid arch is connected with the lower end of the hyomandibular ; it is itself divided on each side into three pieces, the **epi-, cerato-,** and **hypohyals,** which are united in the ventral middle line by an unpaired bone, the **basihyal** or glossohyal, which gives attachment to the retractor muscles of the arch. Articulated to the posterior border of the cerato- and epihyals are eight **branchiostegal** rays, the uppermost of which occupies much the same position as the suboperculum of other fishes.

3·

25. Very similar in construction to the hyoid arch are the succeeding five **branchial** arches, the upper ends of which curve in beneath the skull, to the base of which they are attached by muscle and ligaments, while the lower ends meet in the floor of the mouth. **Epi-** and **cerato- branchials** form the greater part of each arch, but the upper ends are formed of **pharyngo-branchials**, and the lower of **hypo-branchials**, united by certain unpaired pieces, the **basi-branchials**. (Fig. 9).

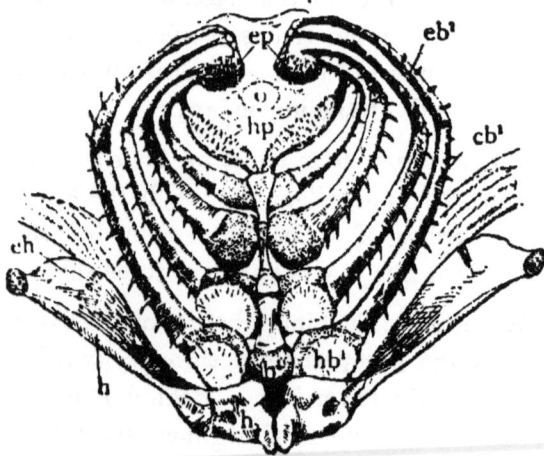

Fig. 9.—Visceral Skeleton of Catfish.

H, hypohyal; ch, ceratohyal; eh, epihyal; i, interhyal; b¹, first basibranchial; hb¹, cb¹, eb¹, hypo- cerato- and epibranchials of first arch; o, œsophagus; ep and hp, epi- and hypopharyngeal tooth plates.

26. The teeth on the premaxillæ and mandible have been referred to above, but the catfish has also a formidable array of teeth further back in the cavity of the mouth. The four plates which carry these are known as the superior and inferior pharyngeal plates, the former of which are attached below the upper ends of the third and fourth arches, while the latter are co-ossified with the ceratobranchials (the only parts present) of the fifth arch.

27. **Appendicular Skeleton.**—The pectoral fin is supported by a bony arch known as the pectoral arch or girdle,

(Fig. 10) which in its turn finds a firm basis of resistance in the skull and vertebral column by means of the supraclavicle or post-temporal described above. This bone is firmly wedged into a socket in the clavicle proper which lies very close to the skin, (especially immediately above the fin where a rough process projecting backwards from it may be felt), and is regarded as a skin-bone, formed on the substructure of the primary arch which it conceals and with which it is closely united. The

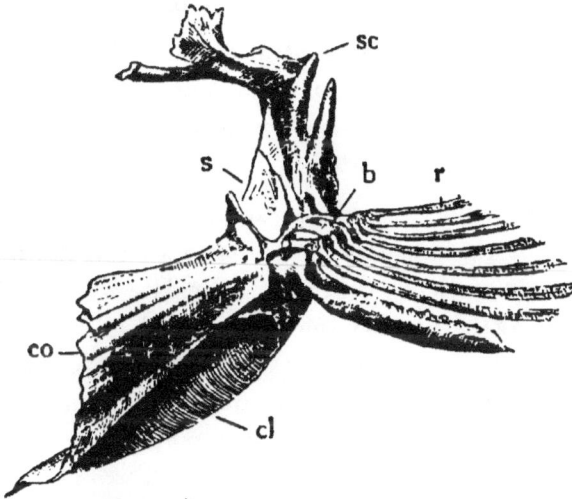

Fig. 10.—Pectoral Girdle of Catfish from behind.

Co, coracoidal, s, scapular portion of primary shoulder-girdle ; cl, clavicular, sc, supra-clavicular portions of secondary shoulder-girdle ; b, basal elements, r, rays of the fin-skeleton.

latter is at first cartilaginous, the former never so. Both the clavicle and the primary shoulder-girdle are divided into upper and lower parts by the fin, the lower parts uniting in the ventral middle line and thus completing the arch. The upper part of the primary arch is known as the **scapula**, the lower as the **coracoid**, and the region where the fin unites with it the **glenoid** region.

2 . Let us now examine the skeleton of the fin itself. It is made up of fin-rays of which the anterior is bony throughout, and toothed on its posterior margin ; it is a **hard** ray or

spine, in contradistinction to the other six rays which are jointed and fringed and are therefore known as **soft** or branched rays. The spine alone is jointed directly to the shoulder-girdle, while the other rays are fixed to it by intermediate pieces of which the hindmost (metapterygial basale) is the largest ; the spine in fact is a ray, plus the foremost intermediate piece, (mesopterygial basale) and their union is to be explained by the use of this ray as a weapon, which may be firmly set (by means of a peculiar joint) and used for offence or defence.

29. The structure of the ventral fin is similar to that of the pectoral. It has eight rays, of which one is hard. There is no pelvic arch, what is generally termed so, being the metapterygial basalia of both fins, united in the middle line.

30. Considerable resemblance to the above will be seen in the fin-rays of the unpaired fins. They are for the most part soft, but some are hard, and they articulate with the **interspinals,** which again fit into the cleft neural and hæmal spines of certain of the vertebræ (§ 13). Of these rays the defensive spine of the dorsal fin deserves mention, as, from the peculiar arrangement of the interspinals with which it is connected, it may be "set" like the pectoral spine.

For the purpose of distinguishing different species of fish it is often desirable to count the number of rays and express them in a formula, (Roman numerals being employed for the hard, Arabic for the soft rays), e. g. for this species :—

$$\left\{ \begin{array}{l} \text{D.} \\ \text{P.} \end{array} \right. \quad \begin{array}{l} \text{I, 6} \\ \text{I, 6} \end{array} \quad \begin{array}{l} \text{A.} \quad 22 \\ \text{V. I, 7} \end{array}$$

31. **Muscular System.**—The muscles of an animal are what we ordinarily call its flesh ; their function is to contract on a stimulus received from the nervous system, and thus to bring nearer together the parts of the skeleton to which they are attached, or to narrow the tubes round which they are disposed. Those surrounding the blood-vessels and intestinal canal form the bulk of what is called the **involuntary muscu-**

lature of the body. They reply slowly to a stimulus, and are not under the influence of the will, while those which unite the various parts of the skeleton are called **voluntary**, because they are controllable by the will, and reply rapidly to a stimulus.

Both kinds of muscles are formed of fibres, which contract on the receipt of a stimulus through motor nerve-fibres which terminate in them. The involuntary are formed of bands of simple muscle-cells— the voluntary of bundles of striated fibres, but in both the muscle-tissue has a framework of connective-tissue which suspends the vessels and nerves distributed to the muscle-tissue.

32. The voluntary muscles are also called "skeletal," for it is obvious that a very important relationship must exist between the skeleton and the muscles. Where muscles are of large size, they must have a sufficient surface for their origin and insertion, and where they cause two parts of the skeleton to move upon each other, the nature and extent of the movements must determine the character of the joint. Thus those parts of the body where the most complicated movements are carried out will have the more differentiated muscles, and those where the movements are simpler, will have the less specialised muscles. In the catfish the more specialised muscles are those which work the jaws, the parts of the visceral skeleton, the gill-cover, and the spines of the pectoral and dorsal fins, while the less specialised are those which form the fl shy mass of the . trunk and tail. The latter exhibit the same metamery which we have seen to characterize this region of the skeleton, for the muscles are divided into **myomeres**, separated by membranous partitions which are attached to the ribs and vertebræ, but the planes of these partitions are not vertical ones, as we may see from a cut through the tail, or from the curved form of the myomeres or fla' es into which the flesh of the fish separates when boiled. Special muscular slips extend into the fins, and serve to depress or erect the rays, but these fin-muscles do not attain the size which the limb-muscles have in higher Ver-

tebrates, where the limbs are of greater importance in connection with the support of the body and with locomotion.

33. Nervous System.—Here we recognize two constituent parts; the central nervous system, and the nerves which course from it to their endings in the various organs of the body, and which indeed are developed as outgrowths from it. The nerves either carry impulses from the central nervous system to the muscles, glands &c., or they transmit impulses to it from the various sense-organs; they are thus either **efferent** or **afferent** in function. They originate from both parts of the nervous system—the brain and spinal cord, and are distinguished therefore as **cranial** and **spinal** nerves.

34. In our study of the skeleton we have seen that the brain and spinal cord are protected in bony canals, perforated for the escape of the nerves. The canals are not entirely filled up by these organs, for certain membranes are present which assist in the protection and nutrition of them. The cranial cavity for example is much larger than necessary to hold the brain, and the interspace is filled up with fatty matter which it is necessary to remove before the brain is exposed. When this is done, it is seen to have the form represented in **Fig. 11** composed of alternately dilated and constricted parts. In front are the **olfactory bulbs** close against the nasal sacs, connecting these with the rest of the brain are the olfactory **tracts,** and then come in order the **cerebral hemispheres,** the **epiphysis** projecting from the concealed **thalamic** region, the **optic lobes,** the **cerebellum** partly covering these and the **medulla oblongata,** with two pairs of secondary swellings upon it. From the under surface will be seen the cerebral hemispheres, with the optic nerves crossing behind them after their descent from the optic lobes, the thalamic region in the form of the **hypophysis** and **inferior lobes,** and the medulla oblongata with the other cranial nerves springing from it.

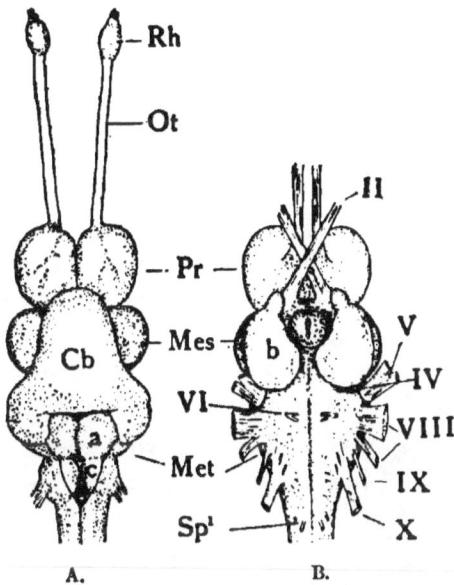

Fig. 11.—Brain of Catfish.

A—From above.　B—From below.

Rh, olfactory lobes ; Ot, olfactory tracts ; Pr, cerebral hemispheres ; Mes, optic lobes ; Cb, cerebellum ; Met, medulla oblongata ; a and c, trigeminal and vagal lobes of medulla ; b, inferior lobes of thalamic region ; II-X, cranial nerves ; Sp¹, first spinal nerves.

35. Transections of the brain show that it is a tube, the walls of which are thick in some parts and thin in others, and the cavity of which is dilated into **ventricles**, communicating with each other. The roof of the cerebral region is so thin that the functions that are discharged by it in the higher animals, must have their seat elsewhere in the fish. A similar condition is observed in the thalamic region, and that of the medulla oblongata, (except anteriorly where the cerebellum is extraordinarily developed), so that it is chiefly the floor of the cavity which is thickened in these parts. On the other hand it is the roof which is thickened to form the optic lobes and the cerebellum.

36. The various regions of the brain or **encephalon** have received the above names as they appear to be comparable to similarly named regions in higher forms ; but the regions and ventricles are also named, by comparative anatomists, **rhin-, pros-, thalam-, mes-, ep-** and **met-encephalon** and the ventricles **rhinocœle, prosocœle, thalamocœle, mesocœle, epicœle, metacœle.** Of these ventricles the thalamocœle is the most complicated as it projects above into the epiphysis, and below into the inferior lobes and hypophysis. The hypophysis and epiphysis are not formed of nervous matter like the rest of the brain, for the former is glandular in its nature, while the latter is supposed to be the rudiment of an unpaired eye.

37. Like the brain, the spinal cord is also a tube, its cavity, the **central canal,** being, however, more uniform, and its walls of similar thickness throughout, except that the side walls are more developed than either the roof or the floor **(Fig. 12).** The metamery of the nervous system is much better seen here than in the brain, for at regular intervals corresponding to the vertebræ, a pair of spinal nerves is attached to the cord. Each nerve originates by two roots, which from their places of origin are known as **dorsal** and **ventral;** the dorsal roots have a knot or **ganglion** upon them, and they contain only efferent fibres, the ventral on the other hand are formed of afferent fibres : both kinds of fibres are, however, soon associated in the mixed nerves of the body.

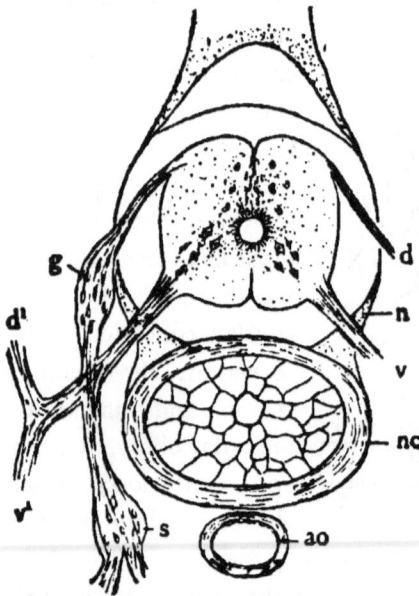

Fig. 12.—Section through Spinal Cord and surrounding parts, in young Catfish. ×40. Nc, notochord, with its sheath ; n, the neural arch ; d, dorsal, v, ventral nerve-root ; g, ganglion of dorsal root ; d¹, v¹, dorsal and ventral mixed nerves ; s, sympathetic ganglion ; ao, aorta.

38. Of the two elements of nerve tissue distinguished in § 33, the nerve cells, also called ganglion cells, are most abundant round the cavities of the brain and spinal cord, but are also found in smaller centres or ganglia, while the fibres are found both in the nerves and in the centres. The function of the fibres is to transmit impulses, and this is effected by the axis cylinder, while the function of the cells is to store up or modify the impulses that arrive through the afferent nerves, and to originate those which are discharged through the efferent nerves.

39. It is easy to understand the way in which the spinal nerves are distributed in the body. Each pair supplies chiefly

the parts of its own metamere, although for purpose of coordination, as e. g., in the various movements of the pectoral fin, the nerves of contiguous metameres communicate with each other in **plexuses**. The parts above the level of the spinal canal are supplied by the dorsal division of the spinal nerves, those below by the ventral divisions, and finally the contents of the cœlom are supplied by special intestinal branches which are provided with ganglia, communicate intimately with each other before supplying the viscera, and constitute the **Sympathetic system**. The arrangement of the cranial nerves is however much more complicated, first, because their metamery is not so evident, and second, because the nerves as they emerge separately from the brain, are not each composed of a dorsal and ventral root, but some seem only to be ventral, others to be composed of several dorsal and ventral roots. Two of them, the 1st and 2nd pairs, olfactory and optic, go to the nose and eyes respectively, the 3rd, 4th and 6th, are motor nerves which control the muscles of the eyeball, the 5th and 7th supply the greater part of the head with sensory and motor nerves, the 8th is distributed only to the internal ear, while the 9th chiefly ends in the 1st gill arch, and the 10th is distributed to the remaining gills, but does not confine itself to the head, and sends branches to the heart, air-bladder and stomach. A separate branch of this widely-distributed (**Vagus**) nerve supplies the sense-organs of the lateral line. The fifth nerve is also not confined to the head, but communicates wi h the dorsal branches of the spinal nerves by a long branch which pierces the back of the skull on either side of the supra-occipital spine.

40. We must now turn our attention to the endings of these nerves, especially to those of the afferent nerves, which transmit impulses to the brain and spinal cord from the sense-organs. Certain of the latter have been already referred to (§ 9), there remain for discussion the higher sense organs, or the nose, eye and ear.

41. Olfactory Organ.—When the roof of the nasal sac is removed (§ 4), the floor will be seen to be formed of a reddish mucous membrane, presenting a median groove and a series of transverse ridges running towards it. A current is established from one nostril to the other, and the odoriferous particles contained, are detected by special olfactory cells, which are situated between the ridges, and are directly connected with the olfactory nerve-fibres. In some fishes the ridges are arranged in such a way as to suggest to anatomists that the nasal sacs are altered gills, which have been confined to this sensitive functions, but the examination of the catfish alone would not suggest this view.

42. The Eye.—In higher Vertebrates the eye affects the shape of the skull considerably more than it does in the catfish, for there is no **orbit** in the latter, and the eye is simply situated in some fatty tissue between the overhanging frontal bone, and the great muscle of the jaw. It is small in size and unprotected by lids, the skin being thin and transparent where it passes over the surface of the bulb, and sufficiently loose to allow the latter some independent movement. This is effected by six muscles, four of which, the straight muscles or **Recti,** are grouped above, below and on either side of the optic nerve, as it courses from the optic foramen to the bulb, and two others, the oblique muscles, cross transversely to the eye from the skull. All the muscles are attached near the equator of the bulb to the **sclerotic** coat.

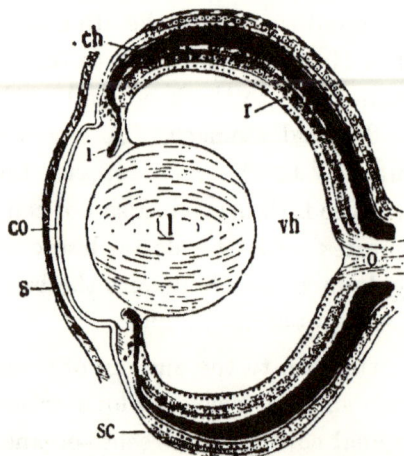

Fig. 13.—Vertical Section of Eye of Catfish. ×30
S, skin; co, cornea; sc, sclerotic; ch, choroid; r, retina; o. entrance of the optic nerve; i, iris; l, lens; vh, vitreous humor.

43. The examination of the bulb itself (Fig. 13) discloses three coats, the outermost of which is subdivided into an anterior transparent part the **cornea,** and a posterior hard fibrous and opaque part the sclerotic. Within this is the second coat, the **choroid,** which chiefly serves to distribute the blood within the bulb, and to form a dark background for the retina, but anteriorly forms a muscular screen—the **iris,** perforated by the **pupil,** through which the amount of light admitted to the sensitive part of the eye may be regulated by the iris. Suspended from the junction of the choroid and iris by a special **ciliary** muscle, is the **lens** a globular transparent body which changes the course of the rays of light admitted to the eye, and casts an inverted image on the **retina** or nervous coat, which lines the whole of the choroid coat, and is separated from the lens by the fluid and transparent **vitreous humor.** The optic nerve which terminates in the retina, must therefore pierce both the sclerotic and choroid coats to do so, and indeed it perforates the retina also, for its fibres form the innermost of the several layers of which the retina is composed.

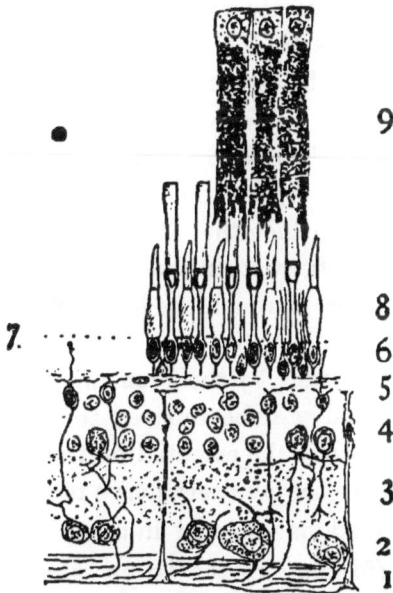

Fig. 14.—Diagrammatic Section of Retina of young Catfish. ×400.

1, layer of optic nerve fibres ; 2, of ganglion cells ; 3, internal molecular, 4, internal granular or ganglionic layer (containing nuclei of Müller's fibres) ; 5, external molecular, 6, external granular layer, containing nuclei of the rods and cones, 8, but separated from them by the external limiting membrane, 7 ; 9, pigmentary retinal epithelium.

44. Study of the development of the eye shows that the retina is really an outgrowth of the brain, which like the brain has a cavity, one wall of which is formed by the single layer of tall columnar cells—the pigmentary epithelium of the retina, while the other, the retina proper, is formed of several layers, of which that toward the cavity is the layer of the rods and cones, while that towards the vitreous humor is the layer of optic nerve fibres. Between the two are various layers of nerve-cells, supported by other elements which are not nervous in their nature. (Fig. 14). The rods and cones are the neuro-epithelial cells of the retina, and the original space between them and the pigmentary epithelium is obliterated by the close contact of the two layers. The lens on the other hand is shown to be developed from the epidermis, and the fibres of which it is composed are really altered epidermal cells.

45. **The Ear.**—In man the ear consists of three parts, the external ear, the middle ear or drum-cavity, and the internal ear or labyrinth : only the latter exists in the catfish. The two former in man are concerned with the concentration of sound-waves on the latter ; how then in the absence of these, do sound-waves reach the labyrinth in the catfish? In some fishes a rudimentary gill-cleft between the hyoid arch and the jaws, appears to be the channel through which the vibrations reach the internal ear, but no such gill-cleft exists in the catfish, so it is probable that they are transmitted through the bones of the head, and above all through the comparatively loose ones, which are suspended to the ear capsule, by the hyomandibular (§ 21). Another possible channel will be referred to afterwards. In all animals the labyrinth is the essential part of the organ of hearing, as it is in it, that the auditory nerve terminates. In most forms the labyrinth is enclosed in a complete cartilaginous or bony capsule, (forming in the latter case the pro*otic*, epi*otic* bones, &c.,) and only perforated by small apertures towards the outside and towards the cavity of the skull, but in the group of fishes to which the catfish belongs, the side of the capsule towards the brain is very deficient, and consequently the greater part of the labyrinth can be seen by opening the cranial cavity.

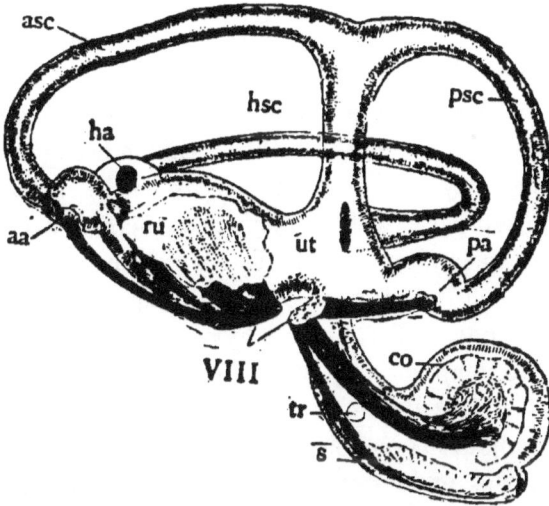

Fig. 15.—Right Ear of Catfish, from within. ˙ ×4.

Asc, hsc, psc, anterior, horizontal and posterior semi-circular canals ; aa, ha, pa, their ampullæ ; ut, utriculus ; ru, recessus utriculi ; VIII, anterior and posterior branches of auditory nerve ; co, lagena cochleæ ; s, sacculus ; tr, opening of transverse duct communicating with the left ear.

46. Of the two parts into which the labyrinth is divided, (Fig. 5) the lower is largely concealed by a little shelf projecting from each ex-occipital bone, and meeting in the middle line over the basi-occipital. This part is known as the **saccule** and **lagena cochleæ**, each of which is a delicate membranous sac, containing fluid (the **endolymph**) and an ear-stone or **otolith** and receiving a considerable part of the auditory nerve, the fibres of which terminate in certain cells of the lining membrane. The upper part is more easily seen ; it is connected with the lower by a narrow duct, and is formed of a central tube, the **utriculus**, with a large membranous sac projecting forwards from it, and containing a very large otolith and corresponding branch of the auditory nerve. Into the utriculus and its recess there open the three **semi-circular canals**, anterior posterior and external, which are respectively situated approximately in sagittal, frontal and horizontal planes. Their lower

openings into the utriculus are dilated into **ampullæ** each of which receives a twig of the auditory nerve.

47. Reference was made to the communication between the air-bladder and the ear of the catfish ; this is brought about in the following way. Near the narrow tube between the upper and lower parts of the labyrinth, there is a cross duct connecting both, and projecting backward into a little pear-shaped sac, which lies in a groove on the upper surface of the basi-occipital bone. The whole labyrinth, and especially this part of it float comparatively freely in the fluid contents of the cranial cavity **(perilymph)**; if currents should be caused in this perilymph it is obvious that currents would also be established in the **endolymph,** and thus excite or disturb the terminal cells of the auditory nerve-fibres which project into it. Such an arrangement for causing currents in the perilymph exists ; it consists in the fact that each half of the neural arch of the first vertebra, can be pressed closer into the neural canal or pulled out by means of a lever (m, Fig. 7), the hinder end of which is attached to the front end of the air-bladder ; consequently any changes of pressure in the air-bladder are transmitted through this lever to the perilymph, and so to the auditory nerve. Whether sound-waves can affect the density of the air in the air-bladder (there is a spot behind the shoulder-blade where it comes immediately underneath the skin) or whether some other function is performed by this singular apparatus, cannot be decided with the knowledge at our disposal.

48. Considerable resemblance will be detected in the microscopical structure of the ends of the nerves of special sense. The labyrinth is lined with epithelial cells which here and there present a different character (neuro-epithelium) where the auditory nerve terminates within them. Two kinds of cells are to be seen, the supporting cells and the hair-cells, the latter alone being connected with the nerves. In the ampullæ the hairs of the hair-cells are very long and delicate, in the other sensitive spots, short and stouter, for they carry on their tips the otoliths referred to above. The ampullæ are supposed to be

concerned in the sense of equilibrium and direction, for which purpose the arrangement in space of the semi-circular canals would appear to fit them.

49. The Intestinal System.—To this belongs the Alimentary canal, with its appendages, the liver, air-bladder, &c.; the gills, which also come under this category, we shall, however, reserve for separate treatment.

We have already studied the osseous boundaries of the gape, and seen the distribution of teeth on these. There are no soft flexible lips, although the skin in this position is more richly provided with the tactile organs described in § 9 than elsewhere. In other respects except in the distribution of pigment and thickness, there is little difference between the **mucous membrane** which lines the mouth-cavity and the external skin. There can hardly be said to be a tongue in the same sense as in the higher vertebrates, but the mucous membrane which clothes the hypo-hyal bones is certainly thicker than it is elsewhere in the mouth. In the living fish the action of the superior and inferior pharyngeal tooth-plates can be seen; the presence of any foreign body causes them to close reflexly on it, and the muscles in connection with them enable them to pass it down into the œsophagus. This is the begining of the alimentary tract proper, in which we recognize three chief divisions, the œsophagus and stomach, the small intestine, and the large intestine. The latter terminates in the anus, in front of the anal fin, and is separated from the small intestine by an **ileo-cœcal** valve, while the small intestine is separated from the stomach by the **pyloric** valve. The limits of the various regions are therefore distinct, but there is no such limit between the œsophagus and the stomach, although considerable structural differences exist between these two organs.

50. The whole intestinal tract is constructed on the same general plan throughout; where it lies free in the cœlom, it is covered by a **serous** coat which is continous with the cœlomic

or **peritoneal** lining by means of a double fold thereof, known as the **mesentery**. If, as in most animals the intestinal canal be longer than the cœlom, it is evident that the mesentery must be longer along the line of its intestinal attachment than along that where it is continuous with the cœlomic lining. If then the intestine be thrown into coils, so as to be accommodated within the cœlom, the mesentery must likewise be complicated in its form. Between the two folds of the mesentery the blood-vessels and nerves which pass to and from the intestine are accommodated. Immediately within the serous coat of the intestine is the muscular coat, in which two layers are recognized, an outer of longitudinal, and an inner of circular fibres. These fibres are of the involuntary order, except in the œsophagus where the inner circular coat is wanting and the longitudinal fibres are surrounded by voluntary or striped fibres. Within the muscular coat we find more or less submucous tissue, answering to the subcutaneous tissue of the skin, and finally the mucous coat which forms the lining of the intestine, and in which, as in the skin, we recognize two layers, a connective-tissue and an epithelial. The latter is the characteristic tissue of the intestine, and forms the bulk of those glands which contribute the various digestive juices. (Fig. 16).

Fig. 16.—Longitudinal Section of Intestinal Wall of Catfish.

Ep, epithelium; m, mucosa; cm, circular, lm, longitudinal muscles.

51. Although there is no marked boundary between the stomach and œsophagus, the former is decidedly wider, and much more abundantly supplied with blood. This is necessary for the proper discharge of the function of the gastric glands, tubes

lined by a continuation of the lining epithelium of the stomach, which dip down into the submucous coat, and consequently make the gastric mucous membrane of considerable thickness. The stomach forms a blind projection beyond the pyloric aperture; it is therefore said to be of the cœcal type, whereas in many other fishes its long axis is directly continued into the intestine. After the food has been subjected to the action of the gastric juice, which renders some of its ingredients more capable of being absorbed, it passes through the pyloric valve (which is simply a local thickening of the circular muscular layer) into the small intestine. Here it is at once mingled with the secretion of two important glands, the **liver** and **pancreas** to be afterwards described, further altered thereby, and finally for the most part absorbed by the walls of the tube and partly propelled further into the large intestine.

52. In most higher animals there are folds or projections of the mucous membrane which facilitate this absorption by the small intestine; except for longitudinal folds, the mucous membrane in the catfish is smooth, and the amount of epithelial surface is increased by tubes projecting into the submucous tissue which are however shorter and wider than the gastric tubes. The chief difference in the structure of the large intestine is in its greater size, and more developed musculature.

53. Of the glandular appendages of the small intestine, the liver is the most important. It is formed for the most part of **hepatic** cells which are continuous with the lining epithelial cells of the intestine through the bile-duct. The cells are suspended in a delicate frame-work of connective tissue, penetrated by blood-vessels, and the liver is therefore very soft in its texture. It has two lobes, a right and a left, each divided into subsidiary lobes, and on the under surface of the right of these is the **gall-bladder**, a reservoir communicating by several ducts with the liver, but only by one with the intestine. Side by side with the

4

last-mentioned duct, is the duct of the pancreas, a gland of different structure and function, which is independent of the liver in higher animals, but in many fishes, is either wholly or partly concealed within it, entering the frame work of the liver beside the **portal vein**, the chief blood vessel of that organ, through which, as we shall hereafter see, a considerable amount of the blood of the body passes on its way to the heart.

54. The liver like the intestine, is within the peritoneum and covered with a serous coat like the intestine, but the **air-bladder** which we have now to examine is only covered by the peritoneum on its ventral surface.   It is therefore an **extra-peritoneal** structure ; it communicates with the hinder end of the œsophagus by a narrow tortuous duct which lies between the folds of the mesentery, and enters the air-bladder a little in front of its middle.   As the air bladder is a recess or diverticulum of the intestine, we should expect to find a similar arrangement of its coats.   As a fact however, the muscular coat is merely represented by a stout white opaque layer, in which only connective and no muscular tissue is present, while this separates with great readiness from the inner mucous coat on account of the scantiness of the submucous tissue.   When the air-bladder is first exposed it appears to be undivided, but when the outer coat is removed there is seen to be a partition subdividing the hinder part into two cavities, and narrowing the apertures by which these communicate with the single anterior cavity.   The inner coat can readily be removed without rupturing it, and then the three communicating compartments are readily seen : it is at the junction of these that the duct enters.   As for the structure of the mucous coat, it is very unlike that of the intestine, for its connective-tissue layer is very delicate, very poorly supplied with blood-vessels, and its epithelial layer, is formed of thin pavement-like hexagonal cells.   Only the outer coat is connected with the altered vertebræ and their processes as described in § 47.

55. As we shall see hereafter, the air-bladder in some fishes is so well supplied with blood, and communicates so freely with the œsophagus that it can act as a breathing organ. Such is obviously not its function in the catfish. Again there are fishes which live in deep water, and can by altering the amount of air in the air-bladder, accommodate their specific gravity to that of the water at any particular level. But such an hydrostatic arrangement must be of less service to a catfish than to many other groups. The connection with the ear renders it likely that the functions discharged by the air bladder are of a complex character, but they are not yet well understood.

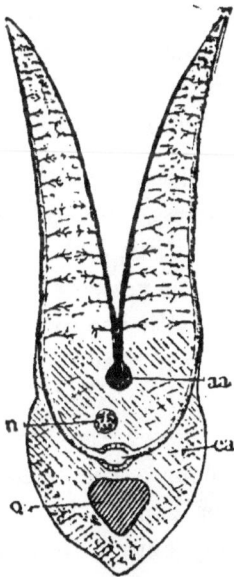

Fig. 17—Diagrammatic Section of Gill-arch.

O, bony arch ; ea, efferent artery; aa, afferent artery; n, nerve.

56. Respiratory System.—We have already studied the skeleton of the gill-arches ; there remain to be examined the soft parts which clothe these. Within the cavity of the mouth there may be observed certain tubercles which fit into each other when the gill clefts are closed, these are the gill-rakers ; they are sometimes of considerable size in other fishes, and may act as strainers of the water which flows out through the clefts, over the gills. On the convex side are the gill-filaments, disposed in two rows. (Fig. 17).

51. The vessels which supply the gill-filaments ascend the arches in a groove, which is easily seen on their convex side in the dry condition. Of the four arches, the last is decidedly the shortest, and the same is true of the slit behind it. All the slits open freely into the branchial cavity, and this by a very wide aperture to the outside, the apertures of the right and left sides being only separated from each other below by a narrow isth-

mus.   In some fishes by the union of the gill-cover to the skin
over the pectoral arch, this aperture may be much reduced in
size.

58. In the roof of the branchial cavity in front of the pectoral girdle,
an organ, the **thymus**, is present in the fishes which attains a considerable
size in some forms, and is difficult to make out in others.   It is seen in the
young catfish, but not in the adult.   Again, below the floor of the mouth-
cavity, round the origin of the vessels which ascend the gill-arches, is
another organ, the **thyroid**, which is of some size in the adult.   The
functions of both of these structures are obscure, but the organs are
found in similar positions in all Vertebrates.

59. **Vascular System.**—In all Vertebrates two subdivisions are present, the blood- and the lymph-vascular system.
The former embraces the heart and blood-vessels and their
contents, the latter the lymph-vessels and spaces and the
lymph-glands in which certain elements of the lymph are formed.

60. Both the circulating fluids contain corpuscles, which in
the blood are of two kinds, coloured and colourless, while in the
lymph only the colourless corpuscles are present (Fig. 3, a). The
coloured corpuscles while passing through the fine vessels of the
gills, have a remarkable power of combining with the oxygen
contained in the water which bathes the gills, but they just as
easily give up this oxygen to the other tissues which require it.
This power they owe to the **hæmoglobin,** which they contain, and which also is the cause of their colour.   The **amœboid** colourless corpuscles have, on the other hand, a faculty
which the coloured ones do not possess to any extent, that of
changing their shape and incorporating foreign particles.

61. The centre of the blood-vascular system is the **heart;**
from it in front are given off the **arteries,** through which the
blood is distributed by way of the gills to the body generally,
and towards it behind, the **veins** converge, through which the
blood is returned to the heart to be again sent on its course.
Between the arteries and the veins are the finer **capillary**

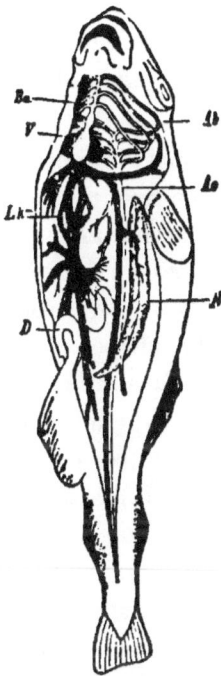

Fig. 18.—Diagram of the Circulation in a Teleost.

(After Claus).

Ba, bulbus arteriosus; Ab, branchial vessels; V, ventricle; Ao, aorta; Lk, capillaries of liver; N, kidney; D, intestine.

vessels, the walls of which are so thin that the blood while flowing through them is enabled to effect exchanges with the surrounding medium. Thus in the gill-capillaries it readily gives up the carbonic acid, which it has accumulated in the tissues, and combines anew with the oxygen in the water, and, while flowing through the capillaries of the rest of the body, gives up to the tissues the food they require, and receives from them the accumulated refuse which has to be removed through the intervention of the glandular cells of the liver and kidney, with which the hepatic and renal capillaries enter into intimate connection. (Fig. 18).

In the catfish the heart is situated between the floor of the hinder part of the mouth-cavity, and the ventral part of the pectoral girdle. A strong partition stretches from the hinder border of the girdle up towards the back bone, and bounds the cœlom anteriorly; it is perforated by the œsophagus and several veins. In front of this partition is the pericardial sac, which contains the heart, and between the partition and the pericardium is the **venous sinus**, to which the various veins converge before they enter the heart. The venous stream is received from the sinus into the **atrium** or **auricle**, a thin-walled chamber, which (when the heart is inspected from below) is largely concealed by the **ventricle**, a thick-walled muscular chamber, which drives the blood through the gills to the body. Connected with the ventricle by a narrow neck is the bulb-like beginning of the great trunk artery, which lies below the copulæ of the gill-arches, and gives off the vessels to these.   These various divisions of the heart are separated from each other by pocket-like semilunar valves, which prevent the reflux of the blood after the contractions of its cavities.

62. The blood-supply of the gills is one of the most interesting parts about the vascular system, because, as we find that the embryos of the air-breathing animals have gill-arches and gill-clefts which disappear as development goes on, so we find that gill-vessels are present also, which afterwards, however, become much altered.   These vessels, partaking of the shape of the parts they accompany, are called the **aortic arches,** and they are uninterrupted tubes which arch on either side from the arterial trunk below the alimentary canal, to another arterial trunk above it, which from its position is called the **dorsal aorta.** Before joining the dorsal aorta, some of the anterior arches give off branches which supply arterial blood to the head. A similar condition obtains in the adult fish, only instead of an uninterrupted aortic arch, there are two vessels to every gill-arch, one which distributes the blood to the gill-filaments, the other which collects it from them.   These are known as the afferent and efferent branchial arteries.   It is therefore the latter which unite to form the dorsal aorta, after the foremost one has on each side given off the **carotid** arteries to the head.   In its course backwards underneath the vertebral column, the dorsal aorta gives off various branches both to the contents of the cœlom, and to the other parts of the trunk and tail, but the venous streams, which collect the blood from these various parts, undergo some delay in their return to the heart. For example, the venous blood from the tail, is in part subjected to the action of the kidney, before it reaches the heart through the **posterior cardinal** veins, (situated immediately underneath the trunk vertebræ,) while the rest of it, with the blood collected from the other contents of the cœlom, enters the liver through the portal vein, and is there subjected to the action of

that gland, before it emerges through the hepatic veins. The venous blood from the head is returned more directly through the anterior cardinal veins, which join the posterior cardinal on each side before they enter the venous sinus.

63. The lymph presents a contrast to the blood in this respect, that it is not contained in well-defined vessels. There are however a series of thin-walled channels, by which the system is put in communication with the cardinal veins. As for the lymph-glands referred to, the most important of these is the **spleen,** a deep red body of considerable size near the stomach, while the second almost equally as large and similar in appearance, but very different in origin, is the **head-kidney,** which lies between the anterior end of the air-bladder and the partition which walls off the pericardial cavity.

64. **Excretory System.**—The structure of the head-kidney is in the embryo catfish similar to that of the kidney proper, which occupies the posterior part of the cœlom, but in the course of growth the excretory tubes which it possesses are replaced by lymphatic tissue, and consequently it has no excretory function in the adult. On the other hand the kidney proper which is separated from the head kidney by the entire length of the air-bladder, is a true excretory gland, which selects by filtration and otherwise from the blood subjected to its action, certain nitrogenous excreta, which have to be removed from the circulation. The excretion is carried off from each half of the kidney, by a separate ureter, the only indication here that the kidney is a paired structure, and that consequently the right and left organs have coalesced. The ureters join on leaving the kidney, and dilate into an urinary bladder before opening exteriorly.

65. There is little external difference between the sexes in the catfish, but there is one notable difference in habit which appears to be common to several allied forms. The eggs after

they have been laid and have begun to develop into the young
fry, are carefully guarded by the male, who keeps them together
and swims over them, returning pertinaciously even after he
has been pushed away. A tropical form, the large eggs of
which are hatched in the mouth-cavity, and a Southern species
allied to our catfish, have been observed to take the fry into the
mouth, and allow them afterwards to escape.

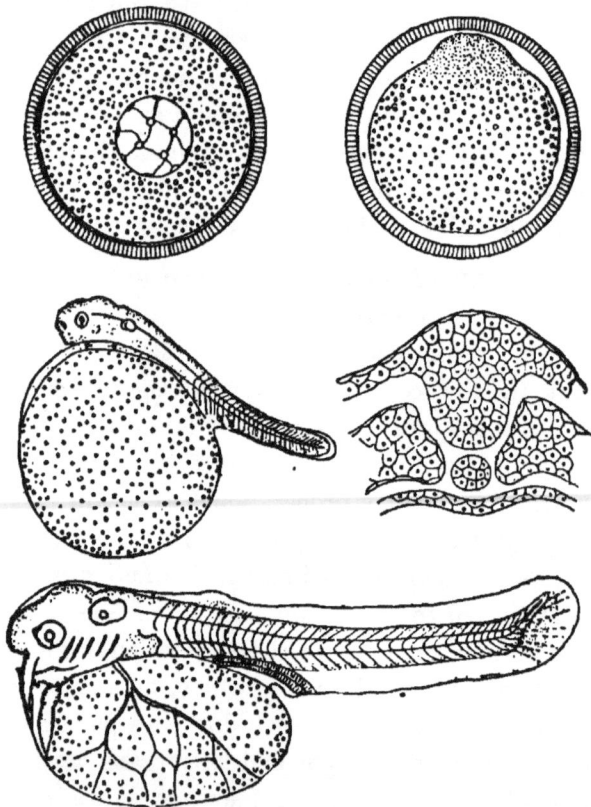

Fig. 19.—Diagram of several stages in Development of Catfish.
(Modified from Ryder).

1, ovarian egg ; 2, egg in which formative yoke has separated to upper pole ; 3, embryo of second day ; 4, section through such an embryo, showing epiblast with nervous system above, hypoblast below, and between them the mesoblast and the notochord ; 5 embryo of sixth day.

Little more need be said about the habits of the catfish. It is remarkable for its tenacity of life, is regarded as a fair food fish, and has accordingly received some attention from pisci-culturists who have found that it prospers also in ponds and streams of other regions besides that in which it naturally occurs.

66. The artifical hatching of fish-spawn with the object of stocking depleted waters and increasing the food supply is now being carried on very vigorously in Canada and the United States, as well as in some European countries. In Ontario the chief hatcheries are at Newcastle and Sandwich, whence vast numbers of the fry of White Fish, Lake Salmon, Pickerel, etc., are distributed for replenishing the waters of the Provinces. The eggs are hatched out in troughs supplied with con-stantly renewed water at a certain temperature, and thus many of the causes which, under ordinary circumstances may lead to the arrest of the development of the eggs are obviated. Some notions as to the gradual development of the body in a catfish may be gathered from Fig. 19.

The egg while still within the ovary of the mother, (1) is about one-eighth of an inch in diameter ; it has two coats, the outer of which is penetrated by minute canals through which the necessary nutriment for the growth of the egg passes inwards. When the egg is laid, the space between the two coats increases in size, and the two constituents of the yolk (the formative yolk which gives rise directly to the body of the embryo and the food-yolk which is utilised as food by the embryo), formerly evenly distributed, now tend to accumulate at opposite poles (2). The formative yolk with its contained nucleus begins to segment, the result being a disc of small cells lying upon the surface of the food-yolk. These cells gradually extend over the whole of the egg, those at the pole arranging themselves into the three layers of the embryo, which already, during the second day assumes a fish-like form (3). It is from the three embryonic layers (epiblast, mesoblast and hypoblast) that all the organs of the body are developed (4) ; a similar arrangement of these exists in all vertebrate animals. The embryo does not escape from the egg membranes until the sixth day, when, although only one-third of an inch in length (5), development has advanced to a considerable extent. Thus the heart is seen in front of the yolk-sac, from the vessels of which it collects the blood enriched by contact with the yolk, and propels it by way of the gills throughout the entire system. The sense-organs, fins, myotomes, and heterocercal tail are all evident, and eventually the yolk is all absorbed and the young fish begins to feed for itself. At the end of three months the form of the adult is attained, although the fish is hardly an inch in length. Teleosts differ very materially as to the length of time of hatching of the egg, and the rapidity with which the developmental processes run, but there is always less food-yolk than in certain other groups of fishes to be afterward referred to, where the development is much slower.

CHAPTER II.

Common Forms of Canadian Fish and their Classification.

1. The common small Catfish, Bull-head or Horned pout which we have been examining is known to zoologists as *Amiurus nebulosus* (Le Sueur). Zoologists as well as Botanists use the Linnæan binomial system of nomenclature, which involves the use of a generic and a specific name for the purpose of indicating to what species any individual animal belongs; of these the generic name stands first, the specific second, and both are followed by the name of the author who first described the kind of animal in question under that specific name, and (if that should be necessary) by the name of the author who first referred the species to its proper genus. The necessity of appending the author's name to a species will be realized when it is understood that two different authors may have described individuals of the same species under different names. Thus *A. nebulosus* has a host of synonymes, one of the most current of which *A. catus* (Linn.) Gill, is given on the assumption that Linnæus had already named our Catfish *Silurus catus*.

It is very hard to find a satisfactory definition for the term "species." In nature we find only individuals; certain groups of individuals resemble each other so closely that we have no hesitation in asserting that they belong to the same species, others may vary so much in colour or in the proportionate size of different organs or in other ways, that zoologists may hesitate whether or no the individuals exhibiting any constantly associated variations should have a separate specific name accorded to them or merely rank as a "variety." The absence of intermediate forms between two or more such groups of individuals is generally considered a sufficient ground for regarding them as distinct species. Varieties are often the result of local conditions and are therefore spoken of as geographical varieties or sub-species, but they may be also brought about artificially by man, in which case they are generally spoken of as " races."

2. Certain groups of species resemble each other so much that they are grouped by naturalists under the same "genus." Some genera are large, embracing a number of species, others small with only one or two ; when they are large it is convenient to arrange the species in smaller groups, which may be designated by sub-generic names. In this way instead of the binomial system, a polynomial system may be adcpted in which the name of an animal may have four parts, the generic, sub-generic, specific, and sub-specific names. Although too cumbrous for general adoption this system has the merit of requiring a close attention to variation, which is one of the most interesting questions in Natural History.

3. In regard to the species we have been studying, the generic name **Amiurus** embraces a large number of different kinds of catfish from different parts of North America. Of these different kinds three occur within our region, viz.: *A. nebulosus*, *A. natalis* (Le Sueur) Jordan, (the yellow catfish), and *A. vulgaris* (Thompson) Nelson, (the long-jawed catfish); some six other species are more southerly forms. The genus is a " difficult " one, the species being hard to characterise and it is doubtful whether all the species are " good." For example there is a southern form which is sometimes regarded as a distinct species [*A. marmoratus* (Holbrook) Jordan], sometimes merely as a mottled variety of our northern form, and consequently named by the zoologists who hold this view *A. nebulosus* var. *marmoratus*.

4. Those characteristics which the individuals of a species possess in common, and which serve to distinguish them from individuals of another species are expressed together with th ir habitat, or range of geographical distribution, in a specific diagnosis; as examples the following diagnoses of the species which occur in Ontario may be copied from a recognized authority :—

A. nebulosus (Le Sueur) Gill.

Colour dark yellowish-brown, more or less clouded, sometimes yellowish, sometimes nearly black. Body rather elongate ; depth contained 4 times in length (measured to the base of the caudal fin). Anal fin usually with 21 or 22 rays, its base contained 4 times in the length of the body. Dorsal fin inserted rather near the adipose than the end of the snout. Upper jaw usually longer than lower. Humeral process more than half the length of the pectoral spine. Length 18 inches. Great Lakes, Ohio Valley, and Eastward. The common bullhead or horned pout of the North and East, abundant in every pond and stream, also introduced into the rivers of California, where it has rapidly multiplied.

A. natalis (Le Sueur) Jordan.

Yellowish, greenish or blackish. Body more or less short and chubby, sometimes extremely obese (var. *natalis*), sometimes more elongate (var. *lividus*). Head short and broad. Mouth wide, the jaws equal (var. *lividus*), or the upper jaw longest (var. *cupreus*). Anal rays 24-27. Great Lakes to Virginia and Texas ; generally abundant. Extremely variable, and running into several varieties.

A. vulgaris (Thompson) Nelson.

Dark reddish-brown or blackish. Body moderately elongate ; depth 4-5 in length. Head 3-4. Barbel long. Mouth wide. Head longer than broad, rather narrowed forward. Profile rather steep, pretty evenly convex. Dorsal regions more or less elevated. Lower jaw strongly projecting. Anal rays 20. Length 18 inches. Vermont to Minnesota and southward ; rather common.

It will be observed that the distinguishing features of these three species are to be found in the shape of the body, the length of the anal fin, and the relative length of the jaws.

5. The generic, specific, and varietal names are generally Latin or latinized Greek words in form, the generic name being always a noun, and the specific and varietal names either adjectives agreeing therewith, or nouns in apposition. Although these names often refer to some characteristic of the form designated, yet this is not always the case, and there is no definite understanding among zoologists as to the principle on which such names shall be selected. For example the name **Amiurus** has been formed to express the fact that the tail-fin is not notched in this genus, whereas the name of the most nearly allied genus **Ictalurus** or **Ichthælurus**, where the tail is notched, is simply a Greek translation of Catfish. Again the specific adjectival name *nebulosus* is formed to express the peculiar clouded colouration of our Catfish, while its synonym *catus*, a noun in apposition with Amiurus is another reference to its common name; further, the varietal names *marmoratus, lividus, cupreus* are adjectives expressing some colour-peculiarity of the forms designated, while the specific name of our great fork-tailed Catfish *I. lacustris* refers to its occurrence in large bodies of water.

6. The species last referred to attains a large size, reaching occasionally a weight of 100 pounds; it is abundant in the Great Lakes and St. Lawrence, and is much used for food. Another allied form the Channel Cat (*I. punctatus*) is found in the channels of the large streams but does not reach the size of the great Catfish. In contrast to these are several species of **Noturus**—Stone-Cats, (Fig. 20) which are rarely more than 4 or 5 inches in length, have the habit of lurking under stones, and are marked by the long adipose fin which is almost continuous with the tail-fin.

7. These three genera are the representatives in Canada of a very large group or "family" of Fishes, the members of which abound in the fresh waters of the tropics of the old and new world, but are only represented in Europe by one species the

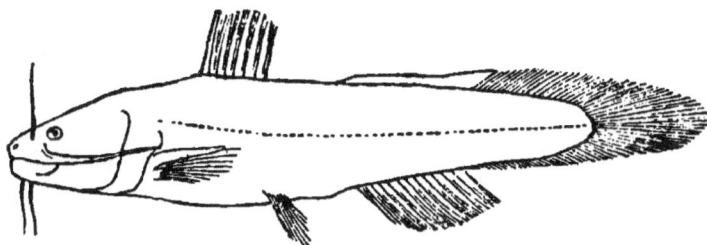

Fig. 20.—Stone Cat. *Noturus gyrinus.*
(After Jordan).

Sheat-fish (*Silurus glanis*) of the Danube. By adding the patronymic ending "**idae**" to the stem of the generic name of this fish, the family name **Siluridae** is formed which thus includes all the genera of the group. Family names are generally formed in this way from some typical or well known genus, and if it is considered desirable to arrange the family into sub-families the termination "*ina*" is generally employed for such smaller groups. But, as in the case of species and genera, zoologists are by no means at one in recognizing the same limits to the classificatory sub-divisions employed. Nevertheless the divisions of various rank are always sub-ordinated to each other in a definite way, thus each of the great primary divisions of the Animal Kingdom or Sub-Kingdoms is divided into Classes, each Class into (Sub-Classes and) Orders, each Order into (Sub-Orders and) Families, while each Family (sometimes sub-divided into Sub-Families or Tribes) includes one or more genera according as the species belonging to it are more or less nearly allied to each other.

8. Thus the Siluroid family as generally understood is one of the largest of the class **Pisces**; it is also a very heterogeneous one, embracing such different forms as our Catfish and the curious mailed Siluroids of the South American Rivers, so that certain authorities consider it to have the rank of an Order (**Nematognathi**), and it certainly does contain forms which are

structurally far more diverse than is the case in other Families of Fishes.   All of its members possess the barbels, the well developed premaxillaries, and the rudimentary maxillary bones of the Catfish, they lack the sub-operculum, but they all have the curiously modified anterior vertebræ and air-bladder, although sometimes these are difficult to detect.   They are for the most part fresh-water forms, but a few are marine.

As we have studied a representative of this group in detail, some account of its most striking tropical forms may be of interest.

Reference has been only made to the fresh-water catfishes above ; there are, however, representatives of the family on the sea coast extending from Cape Cod southwards.   These belong to two genera, **Arius** and **Ælurichthys**, which agree in having the head armed above with bony shields; in this respect they are less like our catfish than the large catfish of the Nile (**Bagrus**) and of the South American rivers (**Pimelodus**) are, and they more nearly approach certain other South American forms— **Doras** and its allies, where the head is completely mailed, but where the branchial aperture is reduced to a mere slit so that water can be retained in the gill-cavity.   This latter condition also occurs in the Electric Catfish of the Nile (**Malapterurus**, Fig. 21) which has no exoskeleton, but has the superficial layer of muscles converted into an electric organ.

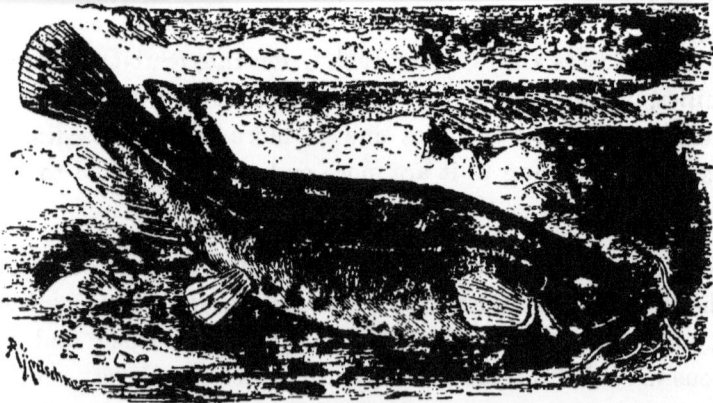

Fig. 21.—Electric Catfish of the Nile.   *M lapterurus electricus.* ⅟₁.

(After Brehm).

A great many of the South American Siluroids have a very complete exoskeleton. **Callichthys** and **Loricaria** (Fig. 22) are representative genera ; they all appear to be very tenacious of life out of water, their gill-cavities being arranged as in Doras. **Aspredo** is a singular genus in which the female carries about the eggs attached to papillæ of the skin of the ventral surface, until they are hatched.

Certain old-world tropical forms are provided with arrangements better adapted than those referred to above for living out of water. **Clarias, Saccobranchus** and others have a recess projecting backwards from each gill-cavity which can be filled with water.

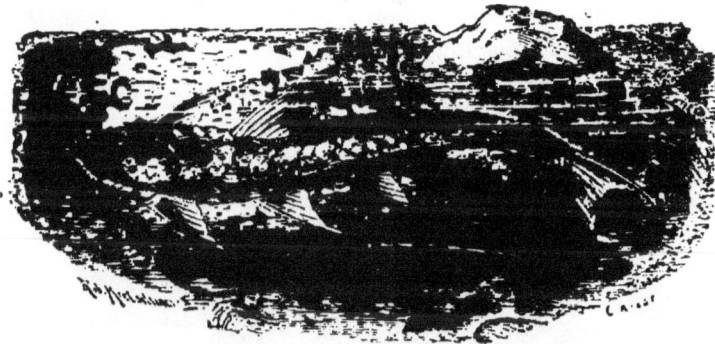

Fig. 22.—Mailed Siluroid, from South America. *Loricaria cataphracta.* ⅓. (After Brehm.)

9. A very large number of our fresh-water fishes belong to a family nearly allied to the Siluroids, that of the **Cyprinidae**, embracing the suckers, carps, goldfish, minnows, shiners, etc., of which the suckers are sometimes reckoned as an independent family (**Catostomidae**). Although very different externally from the Siluroids (for they are generally scaled fishes and often brilliantly coloured), yet they share the peculiar struc ure of the anterior vertebræ and air-bladder, which is present in that group. The gill-cover has all the four bones, but there is no adipose fin. There are no teeth on the jaws, but the pharyngeal bones are well provided therewith. On the roof of the mouth in front of the first gill there is a rudimentary fifth gill called a pseudobranch, through which only **arterial blood**

5

circulates; its meaning will afterwards be explained. This
family is characteristically a fresh-water one, the section of the
Catosomidae being nearly confined to North America, and including our suckers, lake-mullet (Fig. 23), carp and carp-suckers,
while the rest of the Cyprinoids are abundantly represented in
the Old World as well as the New. The suckers and their allies
attain a large size, but the rest of the group are small and very
similar in form and colour, so that they are difficult to diagnose,
and much remains to be found out as to their distribution in
Canada. Two further peculiarities of the family may be referred to, the bright colouring of the males at breeding time in
the spring, and the division of the air-bladder into two or three
compartments by transverse constrictions.

Fig. 23.—The Red Horse. *Moxostoma macrolepidotum.* ⅓.
(U. S. Fish Commission.)

10. The Siluroids and Cyprinoids, like several other families of Teleostei, have an open duct between the air-bladder
and the œsophagus; all the families which possess this are
known as the Physotomous Teleosts, while those in which the
air-duct is absent are known as the Physoclystous Teleosts. We
shall find some familiar forms among the remaining **Physostomi**,
which have the anterior vertebræ separate from each other and
unconnected with the air-bladder. Foremost in importance,
from an economical point of view, is the family of the **Salmonidæ**, which contains so many valuable food-fishes. Chief among
these is the Atlantic salmon, (*Salmo salar*, Linn.) (Fig. 24)

which ascends rivers on both sides of the Atlantic for the purpose of spawning, and is consequently described as migratory
or anadromous, although it is able to live also permanently in
fresh water. Such "land-locked" salmon used to be abundant
in Lake Ontario.

Fig. 24.—The Atlantic Salmon.   *Salmo salar.* ⅟₁.
(U. S. F. C.).

The Pacific salmon, which are canned in enormous quantities in British Columbia, belong to an allied genus **Onchorhynchus**.

11. Our common Lake Trout and Brook Trout belong to a
genus **Salvelinus** differing from Salmo in the absence of teeth
on the vomer; the former (*S. namaycush*) attains a large size,
and is abundant in the larger lakes, the latter (*S. fontinalis*) is
found in ponds and streams, and is well known by its brilliant
colouring, except in those individuals which have access to the
sea, and which replace the red spots and dark bars by an uniform silvery dress. Hardly less important are the common
Lake species of White-fish (*Coregonus clupeiformis*) and Lake
Herring or Ciscoes (*C. artedi*) (Fig. 25) which differ from the

Fig. 25.—The Cisco, or Lake Herring.   *Coregonus artedi.* ⅕.
(U. S. F. C.)

Salmons by their large scales and toothless jaws. Both Salmon and White-fish have certain appendages of the intestines which are not present either in the Siluroids or Cyprinoids; these are situated at the junction of the stomach and intestine and are known as the **pyloric cœca.** They serve to increase the digesting and absorbing surface of the intestines. They are small and few in number in certain smaller marine Salmonidæ such as the Capelin (*Mallotus villosus*) (Fig. 26) and Smelts (*Osmerus mordax*) (Fig. 27), which in spite of their size are of some importance as food-fishes. All of the Salmonidæ have cycloid scales, a pseudobranch and an adipose fin.

Fig. 26.—The Capelin.   *Mallotus villosus.*   ⅓.
(U. S. F. C.)

Fig. 27.—The Eastern Smelt.   *Osmerus mordax.*   ⅓.
(U. S. F. C.)

12. Another important family is that of the **Clupeidae** or Herrings which are nearly all marine fish with a much compressed body, a serrated abdomen and no adipose fin. The commonest species are the herring, *C. harengus* L. and the Shad, *C. sapidissima*; the former spawn in the sea, the latter ascend rivers to do so. One species of **Clupea** the Alewife (*C. vernalis*)

is land-locked in certain inland-waters, and the same is true of the allied Gizzard-Shad (*Dorosoma cepedianum*) (Fig. 28). The latter is of no value as a food-fish nor is the Moon-eye (**Hyodon**) also allied to the Herring-family, but the Anchovies (**Engraulus**) are much esteemed for their flavour.

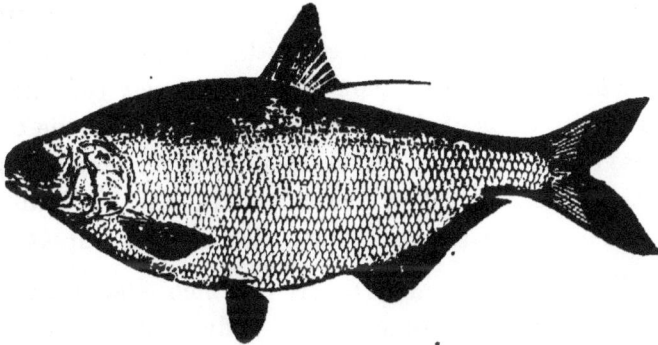

Fig. 28.—The Gizzard-Shad. *Dorosoma cepedianum—var. heterurum.* ½.
(U. S. F. C.)

13. More familiar to inland residents is the family **Esocidae**, a group which is found in the fresh waters of the northern parts of both hemispheres. The largest representative is the Muskallunge, *Esox nobilior*, Thompson (Fig. 29), but the com-

Fig. 29.—The Muskellunge. *Esox nobilior.* ⅟₁₆.
(U. S. F. C.)

mon species is the Pike, *E. lucius*, which is marked with light spots on a darker ground. All the species are voracious and are provided with a large mouth armed with strong teeth. The body is slender and elongated, there is no adipose fin, the pseudo-branchs are concealed and there are no pyloric cœca. As allied to the Esocidae may be mentioned the **Umbridæ**, one species of which, the little mud-minnow (*Umbra limi*), is to be found widely distributed in muddy ditches. The mud-minnow is in some respects allied to the Blind-Fishes of the Southern States (**Amblyopsidæ**), the best known of which and the largest is *A. spelæus* from the subterranean waters of the Mammoth Cave, Kentucky. In accordance with the subterranean life the Blind-fish is colourless, the eyes are extremely rudimentary, but the sensitiveness of the skin is increased by tactile organs like those described in § 9, which are elevated on rows of papillæ above the general level of the skin. In all this family the intestine turns forward and opens underneath the throat.

14. The last Physostomous family to which reference need be made is that of the **Anguillidæ**, chiefly marine forms, but represented in our inland waters by the common Eel (*Anguilla rostrata*). The absence of ventral fins, the confluence of the unpaired fins round the tail, the absence of hard rays, and the rudimentary scales embedded in the soft skin, are some of the chief superficial peculiarities of the group. Allied families are those to which the Bengal **Amphipnous** belongs, which has an air-sac communicating with the gill-cavity, and the Brazilian **Gymnotus** or Electric Eel, in which as in **Malapterurus** the muscles of the tail are converted into an electric organ.

15. **Physoclysti.**—In this division of Teleosts the ventral fins are usually in the course of their development shifted forward till they attain a position either beside or in front of the pectoral fins : they are then said to be **thoracic** or **jugular.** The unpaired fins have generally hard rays. This is especially the case in the **Acanthopteri,** the largest of the orders of Teleosts, to which a vast number of marine forms belong, but which are also represented in fresh-water by the bass and perch tribes. Although the air-bladder never communicates with the alimentary canal in the adult, yet it is developed from it in the

young, and is only afterward closed. It is filled with air from certain blood-vessels which are so arranged as to allow an interchange between the gases of the blood and those of the air-bladder. However great the difference in this way between Physostomous and Physoclystous Teleosts, yet there are some of the latter which are transitional in other respects. The family of the **Scomberesocidae** recalls both the Pikes (Esocidae) and the Mackerels (Scombridae). In addition to the long-billed marine Gar-fishes, several interesting species of Flying-fish, *Exocoetus* (Fig. 30), belong to it, marked by the great size of the pectoral (and ventral) fins. These creatures throw themselves out of the water by means of the strong muscles of the tail, and sustain themselves in the air by spreading the fins.

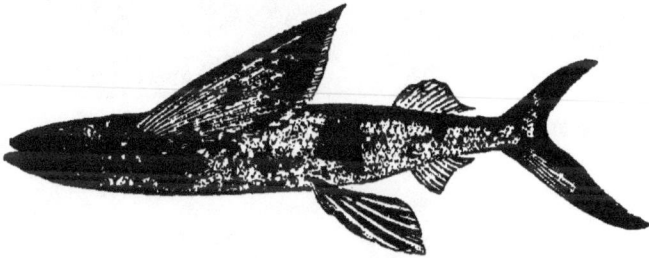

Fig. 3 .—California Flying Fish. *Exocœtus Californiensis.* ½.
(U. S. F. C.)

16. Before discussing the typical Physoclystous fishes a passing reference may be made to certain aberrant forms which attract attention by the peculiarity of their shape. The Pipe fishes (**Syngnathus**), and Sea-horses (**Hippocampus**) (Fig. 31), agree with each other in the structure of their gill-filaments, which are arranged in tufts (**Lophobranchii**), like the teeth of a comb. The snout is much produced, the mouth toothless and the gill-cover a single plate. The Tobacco-pipe fishes (**Fistularia**) have the ordinary gill-structure, but share the elongated body and produced snout of the pipe-fishes. Allied to them are the Sticklebacks (**Gasterosteidæ**) of fresh and brackish waters (Fig. 32), a group of tiny pugnacious fishes which live on the fry of larger fish, but take care of their own in a nest which is constructed and defended by the male. Some of the species have regular bony plates on the side of the body: these however, are absent in our common nine-spined and Brook Sticklebacks.

Fig. 31.—Syngnathus (Pipe-Fish) and Hippocampus (Sea-Horse).
(After Brehm).

Fig. 32.—Two-spined Stickleback.   *Gasterosteus aculeatus.*
(U. S. F. C.)

17. The most characteristic group of Acanthopteri in our region is that of the Sunfishes, **Centrarchidæ,** as a type of which family the common Rock Bass, *Ambloplites rupestris* (Fig. 33), may be examined.   It shares the short compressed body of the rest of the family, the mouth is large and well provided with teeth, for all are carnivorous and voracious forms.   The preopercle is serrated, the opercle ends in two flat points.   The dorsal fins are confluent, there being eleven hard rays in front and

ten soft behind (written D, XI, 10), while the anal fin is VI, 10. Like the other members of the family the colouration is somewhat brilliant olive-green, tinged with brassy hues and mottled with darker colours. Ctenoid scales of considerable size clothe the body ; there are 56 of these along the lateral line in this species. The pseudobranchs are small, and the intestine short and provided with 7 pyloric cœca.

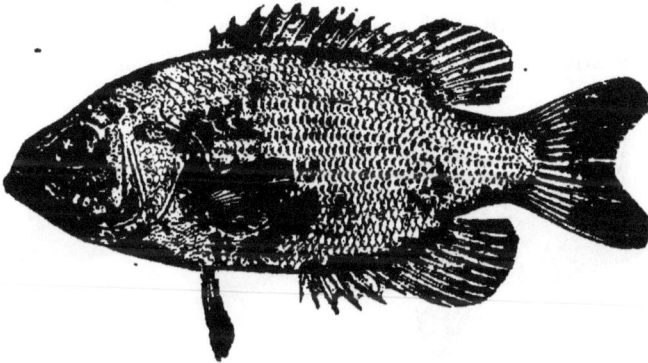

Fig. 33.—The Rock Bass. *Ambloplites rupestris.* ⅓.
(U. S. F. C.)

Besides the Rock Bass several other species of larger size and gamey habits attract the sportsman. These are easily diagnosed by the fin-formula, which for the Grass Bass (*Pomoxys sparoides*) is D.VII, or VIII, 15; A.VI, 17 or 18: for several species of Sunfish, (*Lepomis auritus* and *gibbosus*) D. X, 11 ; A. III, 9: and for the large- and small-mouthed Black Bass (*Micropterus salmoides* and *dolomieu*) D. X, 13 ; A. III, 11. The two latter are well known game-fishes, and apart from the size of the mouth, may be distinguished by the scales of the former being 65-70 along the lateral line and 7-8 in a vertical row above the lateral line, while in the latter they are respectively 72-75 and 10-12.

18. In the nearly allied family of the **Percidæ**, the dorsal fins are not confluent, the anal spines are less numerous, the pseudobranchs are smaller, and the pyloric cœca fewer. Two well marked groups, distinguished alike by size and habit, are

recognized, the **Percina** (including the common perch [*Perca
americana*] and the larger Pike-Perches or Pickerels [**Stizostedium**] as they are generally called in Ontario, and the **Etheostom-
atina** or Darters. The pickerels attain a greater size than the
perch, and are valued food-fishes. The larger species, *S. vitreum*
has three pyloric cœca, its fin formula is D.XIII—I, 21; A.II, 12,
while the smaller *S. canadense* (Fig. 34) has 4-7 cœca, and the
soft dorsal is shorter by three rays. Like the perch the pickerels have a good deal of brilliant yellow colouring on the sides
and are called "dorées" in the Lower Province.

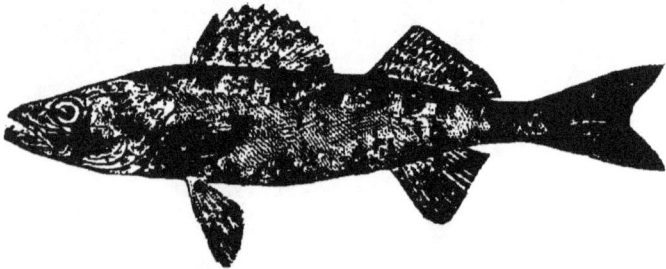

Fig. 34.—The Pickerel. *Stizostedium canadense.* ½.
(U. S. F. C.)

19. Little is known with regard to the distribution of the
Darters in Ontario. They are a characteristic American group,
being amongst the smallest of the Fishes, and distinguished
further by their bright colours, rapid movements, large fins,
and the rudimentary condition of the pseudobranchs and air-
bladder. Some of them conceal themselves under sand, the eyes
alone remaining uncovered, the better to pounce rapidly upon
the active insect larvæ on which they feed. The largest of the
group, the Log-Perch (*Percina caprodes*), attains a length of
6-8 inches, and may be recognized by its black, banded sides,
the smaller forms are the Sand-darter (**Ammocrypta**), Tesselated
darter (**Boleosoma**) and the Striped darter (**Etheostoma**).

20. Among the marine families allied to the above that of
the **Serranidæ** requires mention. It embraces the so called Sea-

bass, one of which, the Striped bass (*Roccus lineatus*), is much valued as a food fish, and is represented in our inland waters by the white bass *R. chrysops* (Fig. 35). Both of them are

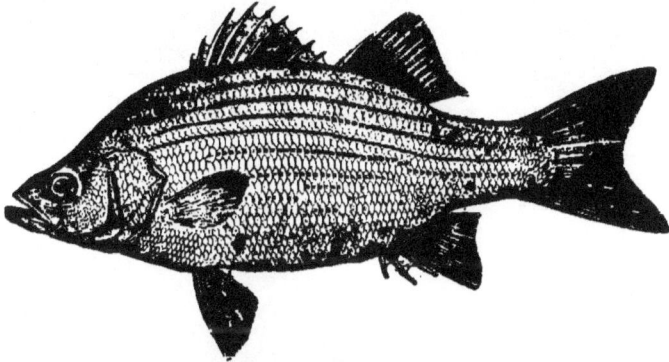

Fig. 35.—The White Bass. *Roccus chrysops*. ½.
(U. S. F. C.)

marked by blackish longitudinal lines, which are more distinct and continuous in the former species. The pseudobranchs are large and the dorsal fins nearly or quite separate.

21. Of the other numerous marine forms of Acanthopteri, the following may be mentioned as of interest, the Mackerel *Scomber scombrus* (Fig. 36) with its numerous dorsal and anal

Fig. 36.—The Mackerel. *Scomber scombrus*. ½.
(U. S. F. C.)

finlets, the Tunny *Orcynus thynnus* (Fig. 37) one of the largest of Teleosts, the Sword-fish (*Xiphias gladius*) (Fig. 38) with its upper jaw prolonged into a sword, and the Sucker (*Echeneis remora*) (Fig. 39) whose dorsal fin is converted into a sucking disc by which the fish attaches itself to moving bodies. A

Fig. 37.--The Horse Mackerel, or Tunny. *Orcynus thynnus.* $\frac{1}{17}$.
(U. S. F. C.)

Fig. 38.—The Sword Fish. *Xiphias gladius.* $\frac{1}{30}$.
(U. S. F. C.)

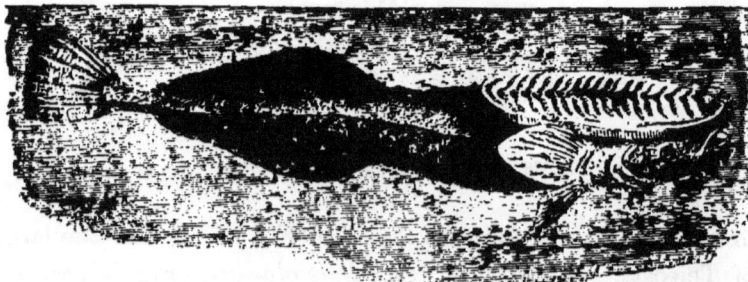

Fig. 39.—Sucking Fish. *Echeneis remora.* $\frac{1}{3}$.
(After Brehm.)

similiar method of adhesion to rocks, &c. occurs in the lump-suckers (**Cyclopterus**) (Fig. 40) where the ventral fins form the centre of the sucking disc.

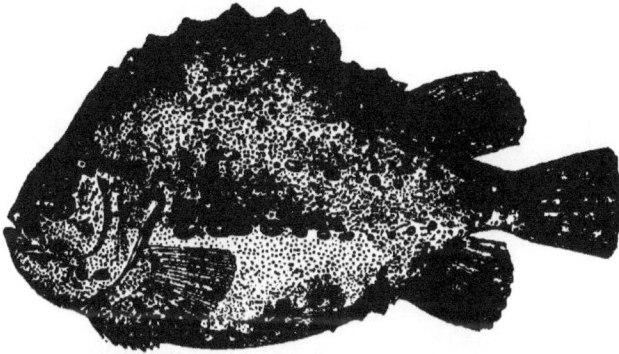

Fig. 40.—Lump Fish. *Cyclopterus lumpus.* ¼.
(U. S. F. C.)

In addition to the Flying-fish above mentioned, one of the Gurnards (**Triglidæ**) possesses the power of flight. (Fig. 41).

Fig. 41.—The Striped Sea Robin. *Prionotus evolans.* ¼.
(U. S. F. C.)

22. Of all the families of Teleosts mentioned, few can compare in economical importance with the **Gadidæ** and **Pleuronectidæ**, which differ very much from each other in their general appearance, but agree in the absence of hard rays from the fins,

whence they are sometimes called **Anacanthini**. They are
marine forms for the most part, the first family including the
Cod-fish and Haddock (*Gadus callarias* and *æglefinus*) with a
host of less important food fishes, and being represented in our
inland waters by the burbot, *Lota maculosa* (Fig. 42). The

Fig. 42.—The Burbot. *Lota maculosa.* ⅓.
(U. S. F. C.)

latter genus has two dorsal fins and one anal, the former three
dorsal and two anal; in both the ventral fins are jugular in
position. To the second family belong the Halibuts, Flounders,
and Soles (Fig. 43) which are generally spoken of as Flat-fish,

Fig. 43.—The Smooth Flounder. *Pleuronectes glaber.* ⅓.
(U. S. F. C.)

from the fact that at an early stage of development they swim
upon one side, which becomes colourless and blind from the eye
moving towards the upper surface of the head.

24. Any account of the Teleostei would be incomplete without a reference to the **Plectognathi**, a group which includes some tropical fish of very bizarre appearance. The File fishes (**Balistes**) (Fig. 44) receive their name from the form of the first dorsal spine, the Trunk fishes (**Ostracion**) (Fig. 45) are enveloped in

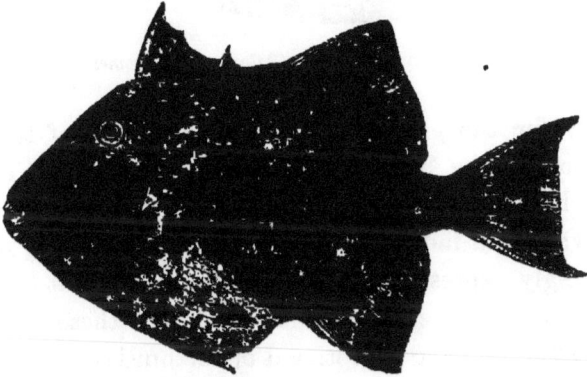

Fig. 44—The File, or Trigger, Fish.   *Balistes capriscus.* ⅓.
(U. S. F. C.)

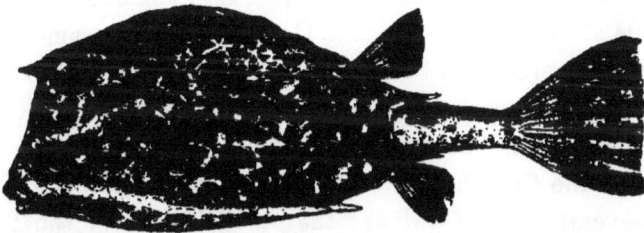

Fig. 45—The Trunk Fish.—*Ostracion quadricornis.* ⅓.
(U. S. F. C.)

a complete box formed of bony plates, the Porcupine-fishes (**Diodon**) (Fig. 46) are covered with long sharp spines, but all agree in the firm union of the bony elements of the jaws to which they owe their name.

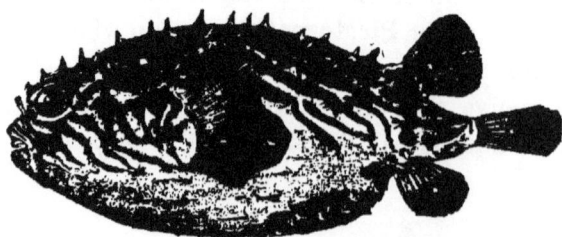

Fig. 46.—The Porcupine Fish. *Chilomycterus geometricus.* ⅓.
(U. S. F. C.)

24. By far the greater number of the Fishes of the present
day belong to the Sub-Class **Teleostei**, but in past geological
times such was not the case, and large numbers of fossil forms
are known which indicate that the other sub-classes, which are
but sparingly represented by living forms, were at one time as
abundant as the Teleosts are now.   One of these Sub-Classes,
the **Ganoidei**, we have exceptional opportunities for studying on
this Continent, because out of the nine genera six are American.
The peculiarities which distinguish the Ganoids from the
Teleosts may be best learned by comparing any one of them
with a catfish, which of all the Physostomi comes nearest to the
Ganoids.   As to its skin the Ganoid is rarely smooth, but
generally covered with bony plates or scales which may be rough
with teeth or smooth with enamel ; the skeleton is cartilaginous
. in the Sturgeon, but as well ossified in the Garpike and Amia
as it is in the Catfish.   The heart has a muscular arterial cone
with several rows of valves ; the pseudobranch of the Teleosts
may be either present as such, or, as in the Sturgeon, as a
functional half-gill on the hyoid arch.   A gill-slit without any
functional gill persists in the form of the "spiracle" in the
Sturgeon between the hyoid and the mandibular arch, and is
more or less complete in the other forms, but the other gill-slits
are concealed as in the Teleosts by a gill-cover.   The air-bladder
opens by a wide duct into the œsophagus and is very richly
supplied with blood, so that in some forms it acts as an accessory

breathing organ. A fold of mucous membrane which serves to increase the internal surface of the intestine and known as the "spiral valve" occurs in all. Finally the vertebral column evidently turns up at the tip in such a way as to divide the caudal fin unequally or heterocercally

The American Ganoids fall naturally into two groups according to the nature of the skeleton; in the **Chondrostei** it is cartilaginous, in the **Holostei** osseous. To the former gro ip belong the two families **Polyodontidæ** and **Acipenseridæ**, of which the former includes the Paddle-fish, *P. spatula* (Fig. 47)

Fig. 47—.The Paddle Fish *Polyodon spatula.* ⅒.
(U. S. F. C.)

of the Mississippi, the latter the ordinary Sturgeon of our Lakes (*Acipenser rubicundus*) (Fig. 48) and the Shovel-nosed

Fig. 48—The Lake Sturgeon. *Acipenser rubicundus.* 1/7.
(U. S. F. C.)

Sturgeon of the Western and Southern States (**Scaphirrhynchops**). In all of these forms the skull is a cartilaginous box adapted to the shape of the brain and sense-organs, and covered with regular bony plates of the same nature as those further back in the body. Polyodon is remarkable for its paddle-shaped snout by means of which it stirs up the mud in the river bottom, on the minute organisms contained in which it feeds. These are sifted out by means of the long and close set gill-rakers which form a very efficient sieve for the muddy water which flows out through the gill-slit. In many respects Polyodon

6

resembles the Sharks in its structure, as indeed do all the Chondrostei, while the Holostei on the other hand approach the Teleosts. The skin is comparatively smooth in the Paddle-fish, but, in the Sturgeon, it is provided with five rows of bony keeled shields, one row on the back and two on the sides, between which shields, the skin is roughened with minute teeth.

25. Of the Holostei we have two genera, each representing a separate family, **Lepidosteus** and **Amia** (Figs. 49 and 50). The

Fig. 49.—The Garpike. *Lepidosteus platystomus.* ⅛.
(U. S. F. C.)

Fig. 50.—The Bowfin, or Mudfish. *Amia calva.* ⅛.
(U. S. F. C.)

latter at first sight looks more like a Teleost, but a closer examination shows the dermal bones of the head, and the unequal division of the tail. The superficial resemblance is chiefly due to the regular rows of cycloid scales, while Lepidosteus is at once marked out by the oblique rows of rhombic enamelled plates, which encase it in a coat of mail. It is to this (and its voracity) that it owes its name of bony-pike, while it is also called garpike on account of the prolongation of both jaws into a beak, a peculiarity present in the marine gar-fishes (§ 81). In many cases where such complete protection is afforded by the exoskeleton, the endoskeleton is incompletely developed; such is not the case in Lepidosteus however, for here all the parts of the latter are completely ossified, the vertebræ being in fact more so than in other fishes, for their bodies are joined together by a ball and socket joint, (the socket behind—opisthocœlous) instead of having a cup at either end as in Amia and the Telcosts. The commonest species is *L. osseus*, but this is replaced in the Southern States by a larger form which reaches a length of eight or ten feet and is known as the Alligator gar.

26. Superficially very unlike the garpike, Amia nevertheless resembles it very closely in internal structure. Its snout is short and rounded, the lower jaws peculiar in being separated by a flat skin bone, the jugular plate, but otherwise the skeleton of the head is very similar to the garpike's.

The dorsal fin is long and low, whereas in the garpike, it is very far back and short and high. The caudal fin is not so unequally divided, and it is marked out in the male by an eyelike spot which stands out against the general dark green hue. There is only one species of Amia, *A. calva*; it is known in different localities by different popular names, among which Mud-fish and Lake Dog-fish are the commonest.

27. In addition to the Ganoid genera already enumerated there are two other living forms confined to the rivers of Africa. These have scales like the garpike and gular plates like the Dog-fish, but their paired fins differ in structure, being composed of a disk-like part containing the skeleton surrounded by a fringe. The commonest species is the *Polypterus bichir* of the

Fig. 51.—*Polypterus bichir.* ⅒.
(After Claus.)

Upper Nile (Fig. 51) the generic name of which refers to the division of the dorsal fin into a series of finlets.

28. Reference was made above to the fact that of all Ganoids
Polyodon is most nearly allied to the Sharks. This is not
merely a superficial resemblance depending on the position of
the mouth but it is seen also in other organs. The gill-arches,
for example, bear between the two rows of filaments, a membranous partition which is hardly present in any of the other
forms but which in the Sharks is much more developed, and
bears the gill filaments. Thus in the Sharks the gill-arches are
not separated by mere gill-slits, but by pouches, the anterior
and posterior walls of which are formed of the aforesaid partitions. The pouches are, at least outwardly, always five in
number (sometimes seven), and they open by a series of slits
uncovered by any operculum (except in the genus Chimaera).
This disposition of the respiratory organs has conferred on the
Sub-Class the name **Elasmobranchii.**

It embraces marine forms familiarly known as Sharks and
Rays ; the former have elongated bodies with the gill-slits on
the sides of the head, the latter are flattened from above downwards, and as broad as they are long, from the enormous development of the pectoral fins, which form the greater part of the
body. Their gill-slits are to be found on the lower surface of
the head. Certain forms are intermediate, in respect to the
size of the pectoral fins, between the Sharks and the Rays ;

Fig. 52.—The Saw Fish. *Pristis pectinatus.* 1/20.
(U. S. F. C.)

the Sawfishes *e.g.*, (**Pristis**) (Fig. 52), which are further distinguished by the enormous development of the rostrum or
snout and the formidable lateral teeth of that organ. Again

Fig. 53.—The Torpedo or Electric Ray. *Torpedo occidentalis.* ⅟₇.
(U. S. F. C.)

the Electric Rays (**Torpedo**) (Fig. 53) are singular in that the muscles of the pectoral fin are largely converted into an electric organ. The Sharks (Fig. 54) are all carnivorous and voracious

Fig. 54.—The Horned Dog Fish. *Squalus acanthias.* ½.
(U. S. F. C.)

forms of great strength and activity; some of the smaller ones (**Mustelus**) live on Shell-fish, but the largest species are often dangerous to man and attain a length of thirty to forty feet (**Carcharodon, Selache**).   In all the Sub-Class the skeleton is cartilaginous, and the skin either smooth or roughened with minute teeth which are similar in structure to the more formidable teeth of the jaws.   They resemble the Ganoids in the structure of the heart and intestine, and in the unequal division of the caudal fin, but they have no air-bladder. The eggs are of large size and are either laid covered by a peculiar horny shell or else the young are born alive.

29. Only another sub-class remains to be discussed, that of the **Dipnoi,** so called on account of the fact that they breathe by the modified air-bladder as well as by gills. Three genera belong here, widely separated geographically, but all similarly situated in that, during the dry season they may have to depend wholly or partly on their lungs for the oxygenation of the blood. They are named Lepidosiren, Protopterus (Fig. 55) and Ceratodus

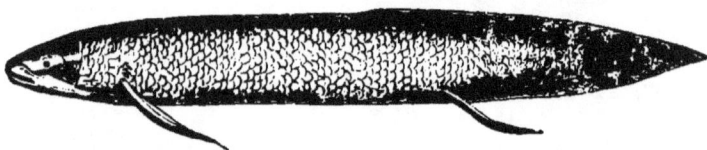

Fig. 55.—The African Lung Fish. *Protopterus annectens.* ‡.
(After Claus.)

and they are found respectively in the Amazons region of South America, on the west coast of Africa, and in Queensland. As regards the structure of the air-bladder they resemble Lepidosteus and Amia, but in respect to the opening place of the air-bladder into the œsophagus as well as to the skeleton and fins, they more nearly resemble Polypterus. The group is of special interest on account of the amphibious habits, and the changes in the respiratory and vascular system rendered necessary thereby. A point in which they resemble the true Amphibia is that the nasal chambers open into the mouth, which is not the case in any of the other fishes.

30. Two aberrant groups of Vertebrates are generally associated with the fishes on account of their fish-like appearance, although in structure they are very unlike them. These are the Lampreys and the Lancelets. The former no doubt owe some of the peculiarities of their structure to their parasitic habits. They attach themselves by their round (**Cyclostomi**)

sucking mouths, which are armed with horny teeth but not supported by jaws, to the bodies of other fish and prey upon them. In general shape they are eel-like : several species are known, some of them marine, but the commonest inland species is the silvery lamprey, *Petromyzon argenteus* (Fig. 56). Apart from the structure of the mouth, they are singular in the respiratory organs, which have seven separate apertures on each side, but only one opening into the gullet. The marine genus **Myxine** has similar habits, but differently arranged gills. In all the forms the skeleton is cartilaginous, and the notochord forms the bulk of the vertebral column.

Fig. 56.—Mouth of River Lamprey.

Petromyzon argenteus.

31. A still further departure from the ordinary vertebrate type is seen in the Lancelets, (**Amphioxus** or **Branchiostoma**) (Fig. 57) little fish-like creatures which burrow in the sand of

Fig 57—Amphioxus lanceolatus.
(After Claus.)

C, oral cirri; ch, notochord; rm, spinal chord; ks, gills; ov, ovary; l, liver: N, kidney; P, branchial pore; A, anus.

the sea coast. They lack the brain, skull, and heart of the vertebrates, but the spinal cord and notochord are present, and the anterior part of the alimentary canal is employed for respiration.

32. The same is true of the marine **Tunicata**, so called on account of the tunic containing cellulose, secreted by the skin around the body They pass through a larval tadpole-like phase, but afterwards lose the tail, and with it the notochord and overlying nervous cord, adopting for the most part a stationary life, during which many of them form colonies by budding.

CHAPTER III.

THE CANADIAN AMPHIBIA OR BATRACHIA.

1. Among the Dipnoi the genus Protopterus is the most truly amphibious, as during part of the year it lives in a torpid condition in the dried mud of the river-beds and is entirely dependent upon its lungs for respiration. It therefore deserves to be called amphibious, far more than do certain members of the Class **Amphibia** or **Batrachia,** such as the common Mudpuppy or Menobranch of our Lakes (*Necturus maculatus*) (Fig. 58), which can live but a very short time out of water. It

Fig. 58.—Canadian Lake Lizard, or Menobranch.
(*Necturus maculatus.*)

is important to compare such a fish-like Batrachian with a true fish, to find out the precise structural differences between the two. The head in the Menobranch is flattened as it is in the catfish, but a distinct neck separates it from the trunk, and here are situated the three pairs of external gills attached to the outside of the corresponding gill-arches and separated by two gill-slits. It has been found that these slits correspond to those between the first and second, and second and third branchial arches in the fish.

No adult fish has external gills of this character, although the embryos of various Elasmobranchs have, and Protopterus has three filamentous gills attached to the pectoral arch.

Perhaps the greatest external difference is in the form of the paired limbs, which no longer resemble the unpaired fin as they do in fishes, but are jointed into the same divisions as they are in the higher Vertebrates, and divided at the ends into fingers and toes, of which there are four to each limb in the Menobranch. But these limbs are not able to support the weight of the body; the chief organ of locomotion is still the tail, and that is flattened and provided with an unpaired fin as in the fishes. It is however not furnished with any skeletal support such as the fin-rays of the fish.

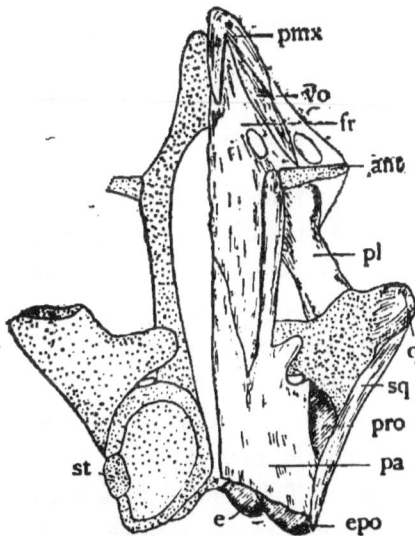

2. At first sight the skin of the Menobranch is very like that of the catfish, but care will be required to make out the system of sensory canals in the head and along the lateral line, and no traces of bony scales will be found in the skin.

Microscopically the most important difference is in the presence of numerous cutaneous glands which furnish the abundant mucus which lubricates the skin.

3. Important differences will be detected in the skeleton, but more in the skull than in the vertebral column. In the latter the individual vertebræ are amphicœlous, and bear short ribs in the trunk region which do not encircle the body cavity. There are no interspinous bones.

Fig. 59.—Skull of Menobranch, from above. (After Huxley.)

Pmx, premaxilla; vo, vomer; pl, palatine; sq, squamosal; pro, prootic; st, stapes; epo, epiotic; fr, frontal; ant, antorbital cartilage; q, quadrate; pa, parietal; e, exoccipital.

Cartilage dotted; cartilage-bones heavily, membrane-bones lightly shaded.

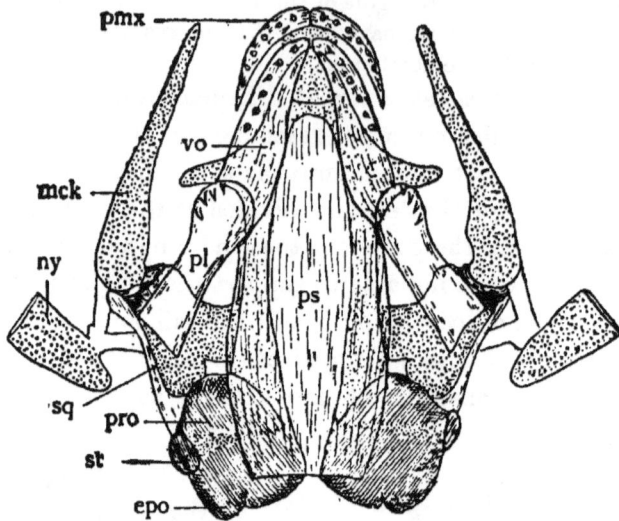

Fig. 60.—Skull of Menobranch, from below.　(After Huxley).

pmx. premaxilla ; vo. vomer ; mck. Meckel's cartilage ; pl. palatine　ps. para
sphenoid ; hy. hyoid ; sq. squamosal ; pro. prootic; st. stapes ; epo. epiotic.

4. The skull retains more cartilage than does that of the catfish, and there are fewer bones to be recognised in it. (Figs. 59
and 60). Of the twenty-seven cranial bones present in the catfish
(I, 17-20), only thirteen are represented, viz.,—paired exoccipitals, epiotics, prootics, frontals, parietals and vomers, and an
unpaired parasphenoid. Some of the other bones are represented
by cartilage such as the mesethmoid and parethmoid ; the nasal
capsule also is a fenestrated cartilaginous capsule, but the
other regions are only membranous. Two new bones are present, the squamosal and the suspensorium on each side, which
lie on the outside of the otic capsule, and thus occupy somewhat
the same position as the pterotic and preoperculum.

As to the jaws, Meckel's cartilage is furnished with a dentary
and an inner spienial bone, and it is hung to the skull by the
Suspensorium, a cartilage which corresponds to the hyomandibular and quadrate of the catfish.　Part of·the hyomandibular

may be represented by a small stapes or columella which fills up a gap or window (the fenestra ovalis) in the outer wall of the auditory capsule, present in all Vertebrates except fishes. A pterygopalatine rod and premaxillæ are present on each side, but the maxilla is even more rudimentary than in the catfish. There is no gill-cover nor branchiostegal rays, but the visceral skeleton is well represented, although the hinder arches are reduced in comparison with the catfish. (Fig. 61).

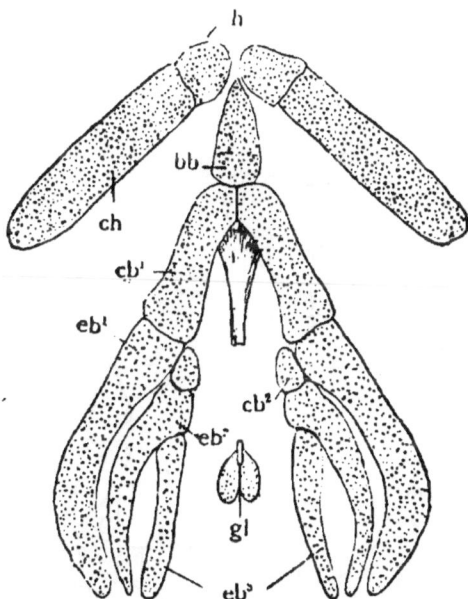

Fig. 61.—Visceral Skeleton of Menobranch.
(After Huxley).

h. hypohyal ; ch. ceratohyal ; bb. first, bb². second basibranchial ; ob¹, cb², first and second ceratobranchials ; eb¹, eb², eb³, 1st, 2nd and 3rd epibranchials ; gl. glottis.

5. Great difficulty will be met with however, in comparing the limbs and the girdles which support them with the corresponding parts in the catfish (Fig. 62). There the coracoids are bony and articulate in the middle line ; here they are cartilaginous and overlap the middle line ; they furthermore each give off a process jutting forward, the precoracoid, some-

times called the clavicle, but a very different structure from the various parts so called in the catfish. Again, the scapula is of much greater size, is ossified and has a cartilaginous leaf-like part above, which nearly reaches its fellow of the opposite side near the middle line of the back. Still the limb is attached to the girdle at the junction of the coracoid and scapula, just as it is in the catfish, only the method of its attachment and the

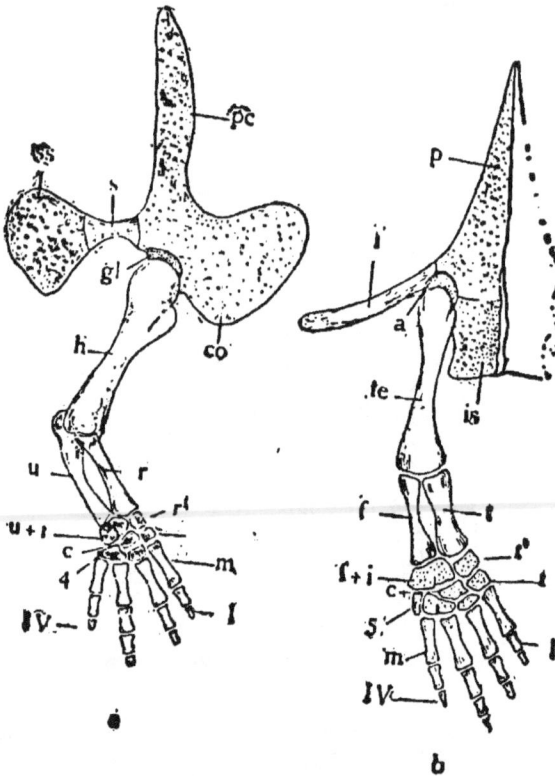

Fig. 62.—a, Skeleton of Anterior, b, of Posterior, Extremities of Menobranch.

pc. precoracoid; s. scapula; ss suprascapula; co. coracoid; gl. glenoid cavity for h, humerus; u. ulna; r. radius; u+i. ulnare + intermedium; r¹. radiale; c. centrale; 1—4, distal carpal row. m. metacarpals; I—IV. digits.

p. pubis; i. ilium; is. ischium; a. acetabulum for fe. femur; f. fibula; t. tibia; f+i, fibulare + intermedium; t. tibiale; c. centrale; 1—5. distal tarsal row; m. metatarsals; I—IV. digits.

nature of its component parts are very different. Only one bone, the **humerus**, effects the attachment and forms the skeleton of the upper arm ; two, the **radius** and **ulna**, are in the fore arm, **carpal** or wrist bones intervene between these and the skeleton of the fingers, which consists in each of a **metacarpal** bone and three **phalanges**. How are we to compare these parts with those in the catfish? It is only possible to do so by studying fishes more primitive in this respect, but investigation appears to show that the humerus is comparable to the metapterygial basal in the catfish, and the other bones to rays in connection therewith. Still greater difference is to be found in the pelvic arch, for what is called so in the catfish is nothing more than the united basals of the fin, whereas in the Menobranch we have a partly cartilaginous and partly osseous pelvic girdle, with the three constituent regions which we shall find in all the higher forms. Of these the uppermost (**ilium**) enters into intimate union on each side with the transverse processes of one of the vertebræ. A similar arrangement exists in all Vertebrates except fishes, and the vertebra (or vertebræ) in question, by which this additional stabili'y of the posterior extremity is secured, is called **sacral**. Corresponding to the glenoid cavity for the attachment of the humerus is the acetabulum for the hip joint. The structure of the skeleton of the hind limb is very similar to that of the fore, and the corresponding parts to those enumerated in the last paragraph are, 1. **femur,** 2. **tibia** and **fibula,** 3. **tarsal** bones, 4. the **metatarsals** and **phalanges.**

6. As the greatest difference in the bony framework is in the limbs, so also the greatest difference in the muscles is to be met with there, but the limbs are of course more differentiated in the higher Amphibia, where they have to perform more complicated duties in connection with support and locomotion.

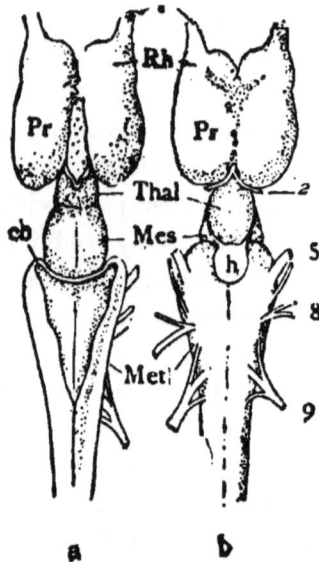

Fig. 63.—Brain of Menobranch.
A. From above.   B. From below.
Rh. Olfactory lobes; Pr. Cerebral hemispheres ; Thal. Thalamic region ; Mes. Optic lobes ; cb. cerebellum ; h. hypophysis; Met. Medulla oblongata ; 2nd, 5th, 8th, 9th. Cranial nerves.

7. In regard to the nervous system of the Menobranch, the most notable difference from the catfish will be found in the brain (Fig. 63). Here the olfactory lobes are united with the larger and thicker-walled cerebral hemispheres ; the inferior lobes are not present, the optic lobes not so distinctly divided into right and left halves, the cerebellum quite rudimentary, and the medulla oblongata destitute of those swellings present in the catfish.

8. The teeth in the Menobranch are not only less numerous, but they are confined to smaller surfaces within the mouth. There are two rows in the upper jaw, of which the posterior (on the vomer and pterygoids) is the longer, while the mandible has only a single row fitting in between the two of the upper jaw. Unlike the fish there is a fleshy tongue free in front and at the sides, and the tubercles on the concave surfaces of the gill-arches are not so prominent as in the catfish. The intestine hardly departs from the tubular form, the liver is more elongated, and the pancreas quite independent and much subdivided.

It is interesting to compare the lungs with the air-bladder of fishes. The glottis is supported by two slips of cartilage which occupy nearly the position of the fourth branchial arches ; it opens into a common chamber whence the thin-walled lungs project backward and two short blind sacs forward ; the latter remind one of the similar points which are present in the air-bladder of Amia.

9. A comparison is more easily effected between the heart and great vessels of the Menobranch, and those of Amia and Lepidosteus or one of the Dipnoi, than those of the catfish, because the ventricle has a muscular cone in these forms which is absent in the catfish. From this the arterial trunk comes off in front (Fig. 64) and divides into two right and left branches, which afterwards subdivide to form the three afferent branchial arteries for the three gills. The blood which is aerated in the foremost and largest gill is partly sent to the head, but partly joins that from the efferent arteries of the second and third gills, so as to form the dorsal aorta. Some of the blood from the second and third aortic arches reaches the lung

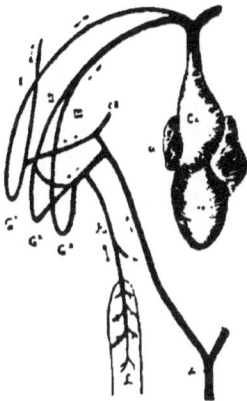

Fig. 64.—Diagram of Heart and Great Vessels of right side of Menobranch.

Co. Arterial cone ; v. ventricle ; a. auricle; i, ii, iii, afferent branchial arteries; G¹, G², G³, the three gills; ce, ci, external & internal carotids ; Pa. Pulmonary artery; o. branch to œsophagus and stomach; L. lung ; D. dorsal aorta.

(only, however, in a partially aerated condition) through the **pulmonary artery,** a modified fourth arch ; its aeration is completed in the lung, whence through a separate vessel, the **pulmonary vein,** it reaches a special compartment of the atrium, not quite separated off from the rest, but partly so by an imperfect partition. In higher Amphibia this partition is perfect, so that the blood within it is not mixed in the ventricle with the returning venous stream, until some of it has been already sent on to the head through the modified first arches **(carotid arteries).**

10. The kidneys are narrow ribbon like structures which extend through the greater part of the length of the body cavity. It is supposed that the Menobranchs spawn in spring, and that they lay eggs nearly of the size of a pea, but further information is desirable as to their habits in this respect.

11. **Class Batrachia or Amphibia.** This class is subdivided into several orders of which three are represented by living forms, the **Urodela, Anura, Gymnophiona,** the others being known merely by their fossil remains. The first order contains the Menobranch and forms allied to it ; the second, the frogs and toads ; and the third, certain tropical earthworm-like forms. It is, therefore, the first two which we have to examine more closely, the ordinal names of which refer to the most striking character, the presence (Urodela) or absence (Anura) of a tail in the adult.

12. Among the nearest Urodelous allies of the Menobranch are some which like it retain their gills throughout life : they are said to be **perennibranchiate** forms, and in this respect are unlike some other Urodeles which lose their gills at a later stage ; these are **caducibranchiate.** Undoubtedly the nearest relative of our **Necturus** is the **Proteus** (*P. anguinus*) (Fig. 65) which is found in underground waters in Carinthia and Carniola. Like the blind-fish of the Mammoth cave it has suffered the almost complete loss of the eyes and the loss of the pigment of the

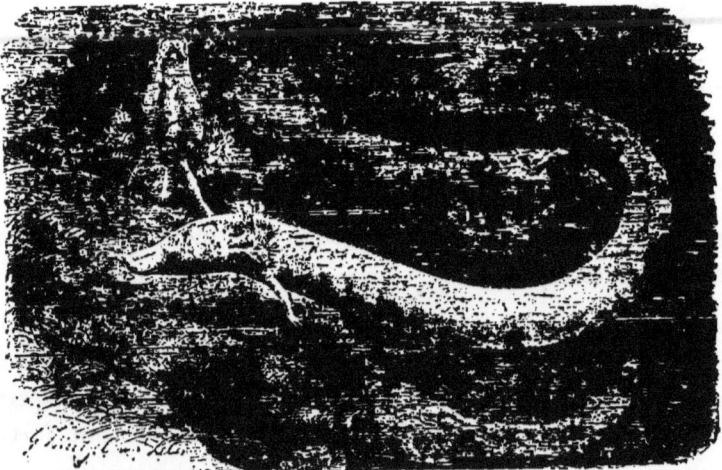

Fig. 65.—Proteus anguinus. (After Brehm).

skin ; the gills, therefore, with the red blood coursing through them, stand out very conspicuously from the colourless body. Instead of the four toes of Necturus, there are three on the front and two on the hind limbs. The only other genus of this group is the **Siren**, of the rice-swamps in the Southern States, *S. lacertina* (Fig. 66) which is eel-like in shape, and lacks the hind limbs. It is less aquatic than either of the other genera, and is able to live out of water for a longer time.

Fig. 66.—Siren lacertina. (After Brehm.)

13. Of the caducibranchiate Urodeles two genera, **Amphiuma** and **Menopoma,** must be regarded as nearest to the foregoing, on account of the fact that in spite of the loss of the gills, one gill-slit on each side (that between the third and fourth gill-arches) persists, whereas in the other forms all trace of these disappears in the course of development. Amphiuma (Fig. 67) is an eel-like form from swamps in the Southern States ; both pairs of legs are present, carrying in one species two, in the other three, toes. Menopoma comes further north, being abundant in the Ohio Valley where it is known as the Hellbender (*M. alleghani-*

7

Fig. 67.—Amphiuma tridactyla. (After Brehm).

ense). It attains a length of two feet and has better developed limbs (with four and five toes) than the foregoing.

A nearly allied form, destitute of the gill-slit, is the giant Salamander of Japan, which grows more than five feet in length.

14. All the other Urodeles are aquatic only in their young stages, and afterwards leave the water for the land where they live either in moist or dry places. As a general rule the tail is rounded in those which have most completely abandoned the aquatic life, in the others it is somewhat compressed. When the new habit of life is adopted, the gills are discarded and all traces of them disappear, the respiration being entirely effected by the lungs. This change, which also involves changes in the vascular system and in the skin, is spoken of as a **metamorphosis**, and it may occur when the creatures are still very small, or it may be postponed till they have attained their adult size, and have even laid eggs. Such is the case *e.g.*, in a large Salamander from Nebraska, *Amblystoma mavortium*, which attains the size of a Menobranch before it loses its gills. It was thought at one time that our Necturus might be such a

larval form, but such is not the case. Another example of arrested metamorphosis is the Mexican Axolotol. A few years ago, this was only known to naturalists in its larval stage, but it has been caused to undergo metamorphosis experimentally, and has been found to do so naturally in some of the localities in which it occurs.

Several Urodeles belonging to this division occur in our region; they belong chiefly to the genera **Amblystoma, Plethodon** and **Diemyctylus.** The largest of these, *Amblystoma punctatum*, the spotted Salamander, attains a length of six inches of which two and one-half belong to the tail. The gills disappear when the creature is two inches long, the colour is purplish black, and each side of the back is ornamented with two rows of bright yellow spots. Of the Plethodons, *P. erythronotus*, the red-backed Salamander, is perhaps the commonest; this species attains about half the size of the foregoing, but loses its gills much earlier than the former does. It lives in moss and under decayed trees where the eggs also are laid. Some allied species are more aquatic in their habits. The newt, eft, or crimson-spotted triton, *Diemyctylus miniatus*, is very common under stones, generally near pools. Its dorsal surface is olive or red, the ventral surface yellow or orange, but the sides are spotted in both varieties with eye-like markings, red with a surrounding black rim.

15. Of the Old World forms allied to these, one of the most striking is the European spotted Salamander (*S. maculosa*) (Fig. 68) which is black with golden yellow blotches. Certain cutaneous glands secrete a milky irritating fluid which appears to be poisonous to small animals. It was thought in ancient times to be most deadly poison, and to have the virtue of extinguishing fire when thrown into its midst.

16. While many Urodela undergo a metamorphosis chiefly characterised by the loss of the gills, the frogs and toads lose at that period not only the gills, but the tail, whence their

Fig. 68—Salamandra maculosa. (After Brehm).

ordinal name **Anura.** Any of the common species of frog will
serve as a type for the recognition of the peculiarities of the
order. As to the general form, the absence of the tail, and the
great development of the hind-legs along with the webbing of
the toes indicate what an entire change in the method of locomotion is to be observed in them. The short plump body also
strongly contrasts with that of the Urodela. Most of the forms
are somewhat brilliantly coloured, and have the power of
altering their colour so as to suit it to the prevailing surrounding
hues. This is not the case in the common toad (*Bufo lentigin-
osus*) which remains concealed generally during the day time,
but it is very marked in the Wood and Green frogs (*Rana temporaria* and *clamitans*), and in the Tree Toad *(Hyla versicolor)*
which, indeed, is very difficult to detect on trees or fences
owing to this faculty. The changes in colour are due to the presence of contractile pigment-cells in the skin which are con-
trolled by the nervous system.

17. It can easily be conceived that the change to a new
medium must be accompanied by changes in the skin. These
chiefly consist in the greater richness of glands which keep the
skin moist and allow it to discharge its subsidiary function as a

respiratory organ, and in the disappearance of those nerve-endings which are only adapted for a watery medium. Accumulations of cutaneous glands are best seen in the toad, behind the ear (**parotoid**) and elsewhere ; they secrete an acrid fluid which must be regarded in the light of a defensive provision. Horny changes in the epidermis or bony plates in the skin, which are common in the Reptiles, are rare in the Anura.

18. As in the Urodela, the skull of the Anura rests upon the vertebral column by two condyles ; it presents in other respects important differences, *e.g.* the girdle-like ossification of the cartilage in the orbital region, and the great reduction of the hyoidean apparatus brought about by the disappearance of the gills. Again, the vertebral column is very much shorter, and its end together with the pelvis have been much modified, in such a way as to offer a solid basis of resistance to the legs in leaping. The shoulder-girdle is very different from that in the Menobranch, chiefly in the median meeting of the precoracoids and coracoids, and the presence of an **episternum** in front, and of a **sternum** behind that symphysis. In the skeleton of the limbs, likewise, we find much change, chiefly in the fusion of the bones of the fore-arm and lower leg, in the great length of the proximal bones of the tarsus, and in the incompleteness of its distal row.

19. In accordance with the gait of the Anura, the musculature of the hind-limb is extraordinarily developed, and the muscles of the trunk are no longer the chief locomotive organs. Apart from the fact that in the frog the brain is somewhat shorter, its olfactory lobes fused, and the optic lobes larger, there is little difference between the central nervous system in the Urodela and Anura. The ear, however, presents a well-marked difference, for there is a **tympanic membrane**, bounding on the outside a tympanic cavity, which communicates with the mouth by an **Eustachian** tube. This tube is comparable to the

spiracle of the sharks and Ganoids, and is closely related to the internal ear even in these forms. A columella longer than that in the Menobranch stretches between the tympanic membrane and the fenestra ovalis.

20. As regards the intestinal apparatus, the Anura present many differences from the Urodeles. The tongue, which is little developed in the latter, becomes in the former the chief organ for securing the insects on which they feed, as it is free behind and can be shot out with great rapidity. It is only absent in two tropical forms, **Aglossa**. The males of some species are furnished with air-sacs, which serve as resonators to reinforce the sounds produced by the larynx, which is better developed than in the Menobranch. Although the adult Anura are carnivorous and their intestine is comparatively short, yet the larvæ or tadpoles have a very long coiled-up intestine. They are omnivorous, but chiefly live on vegetable substances which they gnaw with their temporary horny jaws.

21. It will be at once realized that the metamorphosis of the Anura brings about greater changes both in the form of the body and the habits of life than in the Urodeles. The period of development at which it occurs may be very different, the tadpole phase being sometimes very brief and in other cases much longer. It may in certain cases be retarded by external conditions where it ordinarily occurs early. Most of the forms lay their eggs in water, surrounded by a quantity of gelatinous substance forming the frog's spawn, but other forms which have not free access to water, adopt other plans. In one of the Aglossa for instance, the Surinam toad—**Pipa** (Fig. 69), the eggs are placed in enlarged cutaneous glands on the back of the mother, where they are hatched out and pass through their tadpole-phase. The common toad, again, requires only very small pools in which the larvæ pass their short aquatic life.

Fig. 69.—Pipa Americana. (After Brehm).

Structurally the tadpoles (Fig. 70) differ from the adult chiefly in the presence of the tail, the want of limbs and the nature of the respiratory and circulatory organs. They possess adhesive discs near the mouth by which they attach themselves to aquatic plants and other objects for support. The first gills are external, but these soon disappear, and give place to internal gills on the four gill-arches; these are concealed by a gill-cover, which grows over them in such a way as to leave only a single aperture on the left side. Underneath this gill-cover the fore limbs are first budded out and the hind limbs make their appearance immediately afterwards; both are fully formed before the tail shrivels up. Eventually the gills disappear, and the heart and vessels undergo such an alteration that the venous blood is sent to the lungs and skin to be aerated and the arterial blood to the body generally.

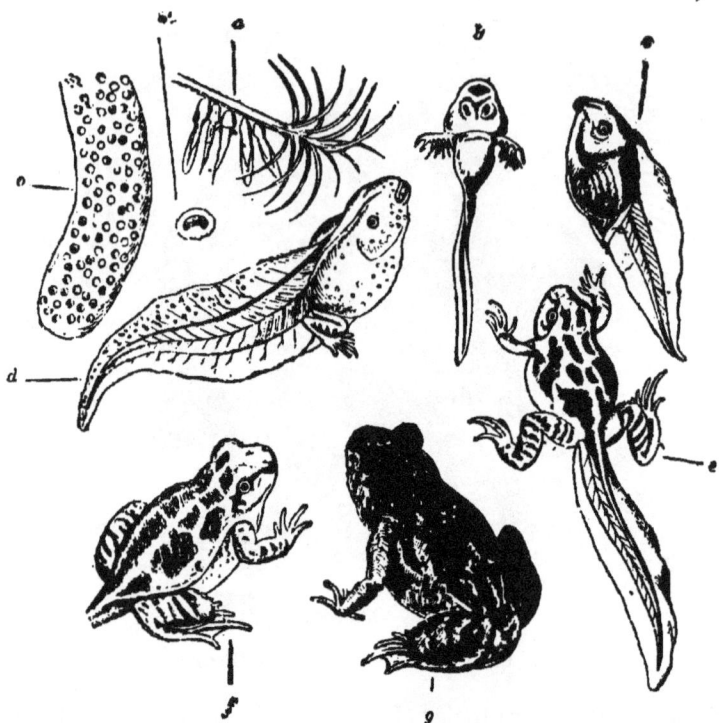

Fig. 70.—Stages in the Development of an European toad. (Pelobates).

o, the gelatinous spawn containing developing eggs, o'; a, a group of tadpoles adhering to a weed ; b, one of these enlarged showing the external gills ; c, stage in which external gills are lost, the spiral intestine is represented ; in d, the hind, and in e, all four limbs are developed, while the tail is shrivelled up in f and lost entirely in g.

22. In III, 16, three genera of Anura are mentioned which form the types of so many families of Anura. The **Ranidæ** are characterized by the great length of the hind limbs, by the presence of teeth in the upper jaw, and by the smooth skin— the **Bufonidæ** on the other hand, have shorter legs, a warty skin and no teeth, while the **Hylidæ** are specially marked out by the adhesive disks with which the fingers and toes are provided, and which permit the climbing habits of the genus. Two species of Hyla, one of Bufo, and five of Rana occur in our region. The largest of the latter is the bull frog (*R. catesbyana*)

one of the most completely aquatic species. *R. clamitans*, the green or spring frog, rarely leaves the water for any distance, while the wood frog *R. temporaria* var. *sylvatica* is found among the fallen leaves of forests, with which its colour is assimilated, while the two remaining species, *R. palustris*, and *halecina* are both found in marshy places and are more variegated in colouration, for there are four or two rows of black spots on the greenish ground of the back.

An interesting case of adaptation to an arboreal life is offered by a species of Ranidæ from the Malay Archipelago—*Rhacophorus Reinhardtii*—(Fig. 71)—in which the webs of the toes are used as a parachute in leaping from tree to tree.

Fig. 71.—Rhacophorus Reinhardtii. (After Brehm).

23. The remaining orders of Batrachia are only represented by fossils from the coal measures and the overlying Permian and Triassic strata. The teeth are generally complex in structure whence the name Labyrinthodontia. In form they resembled the Salamanders, but some attained a gigantic size, and others, such as those found by Sir W. Daw-

son in the hollows of fossil trees in the coal-measures of Nova Scotia, were as small as many of our Salamanders. Many resembled the Ganoids as to the skin-bones of the head, but the notochord was persistent throughout life as in the Dipnoi.

Fig. 72.—Siphonops mexicanus. (After Brehm).

24. The remaining living order, the **Gymnophiona**, is inter-esting, because it embraces forms which, through adoption of a burrowing habit, have undergone the loss of limbs and eyes, and have acquired a hardened skin provided with horny rings. They are represented both in the tropics of the Old and New Worlds (the most northern form is *Siphonops mexicanus*, Fig. 72.), and they appear to be most nearly allied to Amphiuma, which they resemble in depositing necklace-like strings of eggs.

CHAPTER IV.

THE REPTILIA.

1. Our study of the Catfish and Menobranch has taught us that in addition to the aquatic habits which these creatures share, there are certain anatomical features in which they are alike. The Classes Batrachia and Pisces are not separated from each other by any gulf such as meets us when we advance to the study of the Reptilia. It was possible to point out the existence of living forms intermediate in many ways between the Batrachians and Fishes, but we have no such living forms to bridge over the gap between the Batrachia and the Reptiles. Nor do we know any fossil remains which do so ; on the other hand, the Reptiles and Birds, at first sight so entirely unlike each other in structure as well as habits, are, nevertheless, closely allied by fossil forms which present all the important stages of transition between the two groups. Zoologists give expression to these relationships by uniting the Classes Pisces and Batrachia into a group **Ichthyopsida,** and the Classes Reptilia and Aves into a group **Sauropsida.** The proof of the reptilian affinities of the Birds we shall postpone until we have studied the structure of some of our common Reptiles. In the meantime it is necessary to remark that the two Classes Reptilia and Aves are of very unequal rank, as far as the structural characteristics which mark them out are regarded. The Birds constitute a very homogeneous group, the different orders of which are chiefly remarkable for structural features associated with minor differences of habit, but the Reptiles are a very heterogeneous group, and the various families, into which the living orders of **Chelonia** (Turtles), **Lacertilia** (Lizards), **Ophidia**

(Snakes), and **Crocodilia** (Crocodiles) are divided, frequently differ more from each other anatomically, than do the orders of the Birds. The orders themselves, therefore, present still less in common with each other, so that the study of a type of each is necessary to enable the student to grasp the structure of the whole Class. Much attention has been devoted of late to a New Zealand Lizard-like animal, Hatteria, because of all the living Reptiles it is the most primitive form, and most nearly allied to some of the oldest fossil representatives of the Class. Some reference may afterwards be made to this interesting species, but we are obliged to select a more accessible form as an introduction to the group.

2. Although the Turtles in respect to their skeleton are really a very highly-specialised group, yet we shall find in the soft parts many structures which will remind us of the Urodela.

Fig. 73—Snapping Turtle. *Chelydra serpentina*. ⅛.
(After Brehm).

The common Snapper (*Chelydra serpentina*, Fig. 73) one of the least specialised, is a convenient starting-point for the study of the others, but the following description will apply almost equally well to the little painted turtle (*Chrysemys picta*).

The most remarkable point in regard to the skin is the development in it of certain bony plates, which, however, are so closely related to the internal skeleton, as to be more properly dealt with in connection therewith.   In other respects we have to notice here as well as in all the other Sauropsida the great development of the horny layer of the epidermis, no longer confined to the extent of one or two layers of cells, or locally thickened here and there, but developed into the characteristic clothing of scales, scutes, shields or feathers.   In the Snapper we distinguish two kinds of these epidermal appendages, the regular shields which cover the dorsal and ventral surfaces of the trunk, and the smaller and less regular scales and tubercles of the rest of the surface.   In addition to these are the formidable claws with which the distal or ungual phalanges of the digits are provided.   Although all of these structures are formed of horny epidermal cells, yet the underlying mucous cells replace them from below as they are worn off above, and the corium likewise partakes in their formation.   The dorsal surface has three rows of larger carinated shields (five unpaired vertebral, and four paired costal) surrounded by twenty-five smaller ones, of which eleven at each side are called marginal, while that in front is nuchal, and the two behind caudal.   These shields are separated by very scanty connective-tissue from the underlying bones of the exoskeleton.   Similarly, on the ventral surface there are six paired shields named from before backward, gular, postgular, pectoral, abdominal, preanal and anal, of which the abdominal are the largest.   It is these large epidermal shields, which in one of the marine turtles, furnish the tortoise-shell of commerce.   (Fig. 74).   As in the other members of the order, the jaws are provided with horny sheaths like a parrot's beak instead of teeth, and the terminal hooks of these are of considerable size in the Snapper.

After the epidermal shields have been removed, it is seen that their outlines do not correspond to those of the bony

Fig. 74—Tortoise-shell Turtle. *Eretmochelys imbricata.* $\frac{1}{12}$.

plates beneath. The latter belong to the exoskeleton and are arranged in the form of a box, which shelters the greater part of the trunk, and which is divisible into two parts, the dorsal carapace, and the ventral plastron, connected laterally with each other for but a short distance in this species. In the former we recognise a median row, of which the foremost, the nuchal plate, is the largest and is unconnected by bone with the vertebral column, while the eight neural plates which succeed it are co-ossified with the spines of the underlying eight (second to ninth) dorsal vertebræ, and the three pygal plates which terminate the median row, are again free from the vertebræ beneath. Extending laterally from the neural plates are the eight pairs of costals, which overlie the second to ninth ribs, and extend outward for the greater part of the length of these; however, the free tips of the ribs alone are seen to join the bony marginal plates. From this point they are not continued towards the ventral middle line; the bones of the plastron have, therefore, no relation to them.

In contrast to the immobility of the dorsal region of the vertebral column is the great mobility of the cervical and caudal regions. Two vertebræ support by their ribs the ilia, and are consequently spoken of as sacral. With the exception of the mandible and hyoidean apparatus, all the bones of the skull are intimately united, the quadrate is firmly connected with the others, and two bones (which do not occur in the Ichthyopsida), the quadratojugal and the jugal, unite it to the maxilla. The hyoidean apparatus is interesting because it has undergone the reduction which we would have anticipated in an animal where there are no gills. The arches which persist serve for the origin of the muscles of the tongue.

3. As to the appendicular skeleton it has many points of resemblance to that of the Menobranch, the hands and feet are even more primitive, and the shoulder girdle resembles it in the free termination of the coracoids, but the pelvic girdle is somewhat more complicated, presenting a large perforation between the pubis and ischium on each side.

4. The chief points in which the brain of the Snapper differs from that of the Urodela is in the greater development of the cerebral hemispheres and the cerebellum in contrast with the other parts.

5. In the intestinal canal the absence of teeth and the great length of the short intestine are noteworthy, while that we have to do with an air-breathing Vertebrate is sufficiently evident from the large size of the lungs. In most air-breathing animals the change of the air in the lungs is aided by respiratory movements of the thorax, but as the thorax here permits of no such movements, this function is undertaken by the thin wall of the body-cavity in front of and behind the bridges which join the carapace and plastron. The circulatory system is also adapted to the change of the method of breathing, for not only is the blood returned from the lungs into a different chamber from

that returned from the body, but there is also a tendency towards the sub-division of the ventricle, so that the blood which has been aerated is kept towards one side of that chamber, and sent chiefly towards the head.

6. In an important respect the Snapping Turtle resembles the other Sauropsida and differs from most of the Ichthyopsida, viz., the large size of the eggs, which is due to the large quantity of food-yolk present. The eggs are like a bird's, except for the less calcareous shell and the scantier white, and the embryo is gradually developed in it at the expense of the yolk. The period of oviposition is June, some twenty or thirty eggs of the size of a pigeon's being then laid by the mother in a hole scraped out by the hind feet, and not far from a stream or pond. The sun's rays beating on the sandy soil generally selected offer the requisite amount of heat for hatching the eggs out about October.

7. Chelydra is the type of a family named from it which occupies a central position in the order of the Chelonia, and it is easy to proceed from it to the wholly aquatic turtles on the one hand, and the wholly terrestrial forms on the other. At the two extremes are forms which differ very materially from each other in the adaptation of the form of the body to the surroundings. The Marine turtles are very much depressed, their feet are converted into flippers, and the carapace is not adapted for the protection of the retracted head and limbs ; on the other hand, the purely terrestrial forms have a very convex carapace within which the head, tail, and limbs can be sheltered, and in some forms (the box-turtles) the plastron is hinged in such a way as to close effectually the anterior and posterior apertures into the shell.

8. We may first proceed in the direction of the more aquatic forms, of which the soft-shelled turtles, Trionychidæ, are fresh-water animals. There are two common species *Amyda mutica* and *Aspidonectes spinifer*, abundant in streams opening into the great lakes from the South, and in

both the edge of the carapace, and the whole of the plastron are of leathery consistence. These forms lie buried in mud and may remain for hours under water; their respiration is then effected by water taken in and rejected through the nostrils, in such a way that the mucous membrane of the pharynx, which is provided with vascular papillæ arranged on the arches of the visceral skeleton, is constantly bathed with fresh water. This is an interesting point of contact with the Urodela.

9. The feet of the Trionychidæ are broadly webbed, but not converted into flippers as they are in the marine turtles, the Chelonidæ. In these the anterior flippers are largest, and the claws are much reduced. One of the genera, Dermatochelys, has a leathery skin in place of the horny shields which are present in the other genera, the green or edible turtle (*Chelonia mydas*), and the Tortoise-shell turtle (*Eretmochelys imbricata*), (Fig. 74), in which latter form the horny shields overlap each other.

10. Proceeding from Chelydra towards the turtles of more terrestrial habit, in all of which the plastron is much more complete than it is in that genus, we come first to the Cinosternidæ in which the carapace is more vaulted, although the feet are still webbed, the creatures living for the most part in muddy ponds. The most northerly American form is the Musk turtle (*Aromochelys odoratus*), the secretion from the cutaneous glands of which has a somewhat offensive musky odour. Closely allied are certain more southerly mud-turtles, which are able to close the shells.

11. Most of our species of turtles, however, belong to the Emydidæ, all of which are aquatic when young, some like the painted turtle (*Chrysemys picta*), and the spotted turtle (*Nanemys guttata*) throughout life, while others like the Wood turtle (*Chelopus insculptus*) are found in dry places away from water. The most ter-

Fig. 75.—European Land-Tortoise. *Testudo graeca.* ⅓. (After Brehm).

restrial of the family is the common Box-turtle (*Cistudo carinata*), in which the plastron can be shut up over the retracted extremities. It lives in sandy hills, and forms burrows into which it retreats during rain.

8

12. Finally the **Testudinidæ** embrace the truly terrestrial tortoises represented by one species in the Southern States, but occurring abundantly in the warmer parts of the Old and New Worlds. (Fig. 75);

13. The genus Hatteria (Fig. 76), referred to above is most nearly related in its habits and form to the Lizards, **Lacertilia**, but there are some respects in which its structure is much more primitive; *e.g.*, its vertebræ are amphicœlous and its pineal body (I. 36), presents more nearly the structure of an eye than does that of any other living reptile. Unlike the Lizards its quadrate bone is united firmly with the skull, and by an arch below the eye with the maxilla.

Fig. 76.—*Hatteria punctata.* ⅓.
(After Brehm).

14. In spite of the difference in habit between the extreme forms of the Chelonian series, there is not so much difference in external appearance as we meet with in the second order—the Lacertilia. A few aquatic forms belonging to the Varanidæ,

like the large water-lizards of the Nile, do not exhibit any special adaptation for locomotion in water. Most of the forms are terrestrial in their habits while some are arboreal, and others lead a subterranean life. In accordance with such differences in the surroundings, we find great differences of external form. The members of the order are especially abundant towards the tropics, only two families being represented further north by the Blue-tailed Skink (*Eumeces quinquelineatus*) and the Brown Swift or Pine-tree lizard (*Sceloporus undulatus*). Both of these lizards are of small size and very active creatures, the last mentioned belonging to a large family the **Iguanidæ**, which embraces most of the New-World lizards. The forms which lead an active arboreal life are generally compressed in shape, while those which creep about in sandy places depending on their colour for protection, like the Horned Toad of the Southern States (*Phrynosoma cornutum*, Fig. 77), are depressed. Among

Fig. 77—Horned Toad. *Phrynosoma cornutum.*
(After Brehm).

the largest members of the family are the great Iguanas of the Brazilian forests, which are alike remarkable for their size and for the singular crests and combs with which the skin is adorned. An old world family the **Agamidae** contains forms which resemble in habit and appearance some of the Iguanidæ. Thus there is an Australian species (*Moloch*) in which the skin, as in the Horned Toad bristles all over with spines, while again there are many active arboreal forms. Among the most interesting of these is the Flying Lizard (*Draco volitans*) (Fig. 78), a curious little Indian form in which the foremost

Fig. 78—Flying Lizard. *Draco volitans.*

false ribs, which do not reach the breast-bone, project straight out from the body, and have the skin stretched between them in such a way as to form a serviceable flying membrane, which enables them to drop obliquely through the air in their hunt after the insects on which they live.

15. Scarcely less well adapted for an arboreal life are the Chamæleons and the Geckos, the former confined to the Old World, the latter found in the tropics of both Old and New Worlds. In the former the feet are shaped something like those of a climbing bird, the five toes being arranged in opposite groups of twos and threes, the better to grasp the branches on which they perch, while in the latter (Fig. 79), the toes are provided with adhesive discs, which enable them to climb up vertical surfaces such as walls and rocks. Both families are insect-eaters, but the Chamæleons secure their prey by shooting out the long worm-like tongue, while the Geckos spring upon theirs from a distance. While the Chamæleons are strictly arboreal forms and are protected in the foliage in which they live by assimilating their colour to that, the Geckos are also to

Fig. 79—Geckos. *Platydactylus mauritanicus.* ⅓.
(After Brehm).

be found in treeless districts, running over rocks and living in inhabited houses.

16. In several families of Lacertilia on the other hand we meet with a form of body adapted for creeping rapidly through underbrush and underneath stones, as well as for burrowing in the ground. In such creatures the body is cylindrical, almost snake-like, the limbs being either rudimentary or entirely absent. Generally the hind limbs are indicated even when the fore are absent, but in one Mexican genus (*Chirotes*), it is the latter which are alone present. In the Glass

Snake of the Southern States, *Ophiosaurus ventralis*, as
in the European Blind-worm, *Anguis fragilis*, (Fig. 80),
there are no limbs; both of these are extremely fragile

Fig. 80—European Blind-worm. *Anguis fragilis.* ⅓.
(After Brehm).

creatures, the tail being readily cast off in violent efforts to
escape from a capturer. The most completely adapted for an
underground life is the Amphisbæna, (Fig. 81), of South

Fig. 81—*Amphisbœna alba.* ⅓.
fter Brehm).

America. This curious lizard is cylindrical in form, the head
and tail both abruptly rounded off; they live in ants' nests
and feed on their larvæ.

17. From such footless lizards to the true snakes, **Ophidia,** the transition is sufficiently easy. Among the latter, indeed, are some forms which retain rudimentary hind limbs; such are the Pythons, Boa Constrictors and Anacondas of the Old and New World; again there are other smaller forms like the blind snake (*Typhlops*), which show the burrowing habit and the external form of the Amphisbæna, but lack the peculiar arrangement of the jaws which we see in the typical snake. All our Ophidia, however, belong to two families which exhibit considerable difference from the structure of any lizard. Not only are the fore and hind limbs absent, but there is no trace of the girdles supporting them, nor of a sternum. Locomotion, being effected by the ends of the ribs and by shields of the ventral surface whose hinder edges are free, presents a great contrast to the clumsier movements of the footless lizards. In the latter, certain of the organs are affected by the length of the body, the tendency being for paired organs like the lungs and oviducts to become unequal in size, one of them assuming the function of both, but this tendency is carried further in the snake, so that one lung or one oviduct may alone be present.

18. The absénce of the pectoral and pelvic girdles makes it impossible to recognise any but the trunk and caudal regions of the vertebral column. A neck may be present in the form of a constricted part of the trunk behind the head, as in the Rattlesnake, but its vertebræ do not differ from those of the region behind it. So heterogeneous an order is that of the lizards, that we meet with the greatest variety in the epidermal coverings of the body, but the Ophidia constitute just as homogeneous a group on the other hand; nevertheless, in spite of the apparent similarity of the scales and shields, slight differences in the form and arrangement of these are used by systematists in the diagnoses of the species. (Fig. 82). The epidermis is cast off several times a year in the form of a slough, the first moult taking place immediately after the escape from winter quarters.

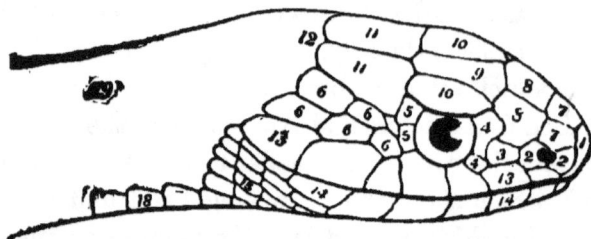

Fig. 82.—Scales of the head in Coluber (Bascanium) constrictor.

(After Garman.)

1, Rostral. 2, Nasals. 3, Loreals. 4, Preoculars or Antorbitals. 5, Postoculars or Postorbitals. 6, Temporals. 7, Internasals. 8, Prefontals. 9, Frontal, 10, Supraciliaries or Supraoculars. 11, Parietals. 12, Occipitals. 13, Labials. 14, Infralabials. (Between the infralabials are the submentals, and clothing the tip of the lower jaw the mental). 18, Ventrals. 19, Dorsals.

19. One of the chief peculiarities of the Ophidia is their method of securing their prey, and bolting it undivided. Some of the larger forms are dependent entirely on the flexibility of the vertebral column and the power of the trunk and intercostal muscles for strangling their prey, but the smaller snakes either seize their victims with their teeth, or first inflict a fatal wound with their poison-fangs. The poisonous snakes must be regarded as the most specialized of the Ophidia, for the teeth are not only reduced in number in comparison with the harmless forms (where they may be as numerous as in the Teleosts), but the poison-fangs (which are confined to the maxillaries), are either provided with a groove on the anterior surface, or with a canal connected with the duct of the poison-gland,—a specialized part of the glands of the upper lip, which is compressed by the muscles which close the jaws. That the snakes may be enabled to swallow their booty whole, a process which is often a very gradual one, the parts of the mouth are provided with extraordinary mobility. The pterygo-palatine bar is capable of greater movement than is even possible in the fishes, where it will be remembered its bones are not incorporated with the cranium, and not only is the quadrate bone freely moveable, but the squamosal which supports it, is hinged to the skull, from which it projects backwards. Thus the articulation of the lower jaw, which is composed of two movable halves, is situated behind the head, and the gape is consequently extremely wide.

20. The Ophidia are destitute of the Eustachian tubes and tympanic cavities, the outer ends of the columellæ merely abutting against the quadrates.

Most of the snakes lay eggs, which are hatched without the aid of the mother, but some of the venomous snakes, as well as fresh-water forms, bring forth their young alive, and these are in certain instances taken care of by the mother.

21. Apart from the narrow-mouthed Typhlops and its allies, and the Pythons, Boas, etc., with rudimentary hind-limbs, the Ophidia fall into three groups, the extremes of which are formed by the poisonous rattle-snakes on the one hand with few canaliculate poison-fangs, and the harmless Colubridæ with numerous non-perforated teeth, on the other, while the various poisonous snakes with grooved teeth, like the brilliantly-coloured Bead-snake of the Southern States (*Elaps*), the spectacled snake of India (*Naja*), and the flat-tailed sea-snakes of tropical seas (*Hydrophis*), occupy an intermediate position. In this region only the extreme forms are represented, the Crotalidæ and the Colubridæ—the former embracing the rattlesnakes and copperheads, the latter all our numerous harmless snakes.

22. *Crotalus horridus*, the banded rattlesnake, is marked by the head being covered with scales instead of regular shields, and by its alternate bands of two shades of brown. As in all the more venomous snakes the head is sharply marked off from the body by a neck. The movements are much more sluggish than in the Colubridæ, the greater agility of which compensates them for the absence of the peculiar weapons of the rattlesnake. One of the most characteristic features of the genus is the rattle formed of singular epidermal scales, the function of which has been much discussed. Observers are not agreed whether it is used to attract prey or to frighten away enemies. It is possibly useful for both

purposes. Snakes which are destitute of a rattle have been observed to make a rustling noise with the tail, and it is interesting in considering the origin of the rattle to recognize that each successive ring is merely the retained slough of the tip of the tail.

By the absence of a rattle, the presence of cephalic shields, and the smaller size, the Copperheads, *Ancistrodon contortrix*, are readily distinguished from the Rattlesnakes, which they resemble, however, in being very venomous. They are found in less rocky ground than the foregoing, are somewhat more active in their habits, but seek similar prey, viz., small animals, birds and frogs.

Of the Colubridæ the Garter Snake, *Eutænia sirtalis*, is certainly the commonest. Its dorsal scales are carinated, and arranged in nineteen rows, while those of the ventral surface of the tail are undivided. An allied species, *E. saurita*, the Swift Garter Snake, is much slenderer, and has a longer tail.

The Garter Snakes affect damp swampy places, take readily to water, and are gregarious in their winter quarters. They are viviparous like most aquatic snakes. The commonest Water Snake is *Tropidonotus sipedon*, which is to be seen basking on the shores of streams, to which it takes when startled. Another common form is *Storeria Dekayi*, the Little Brown Snake. It is also aquatic and insectivorous in its habits ; its dull colours present a strong contrast to the bright green of the Grass Snake, *Cyclophis vernalis*, a form which lives in marshes, and attains a length of eighteen inches. Two larger species, the Black Snake, *Bascanium constrictor*, and the Fox Snake, *Coluber vulpinus*, prey upon larger animals such as mice and frogs, and attack birds' nests. Both of these species attain a length of five or six feet. The one is to be recognized by its uniform black colour, while the other is light brown with darker blotches. Finally the Milk Snake may be mentioned, *Ophibolus triangulus*, a whitish snake with oval brown blotches edged with black, found in dry situations, and visiting dairies for the milk ; the Ring-necked Snake, *Diadophis punctatus*, with its characteristic yellow ring, and lastly the Hog-nosed Snake, *Heterodon platyrhinus*, a peculiar form generally supposed to be venomous, which has the habit of distending its neck with air so as to look formidable, and then emitting the air with a hissing sound, whence it is also called Blowing Viper. In the poisonous genus Naja, a similar formidable appearance is secured by the stretching out of the foremost free ribs at right angles to the vertebral column, so that the neck is converted into a flattened disc.

23. The fourth and last order of living Reptiles is that of the Crocodilia, aquatic forms of large size which are found in tropical rivers over the whole world. There are three families represented by the Gavial of the Ganges (Fig. 83), characterized

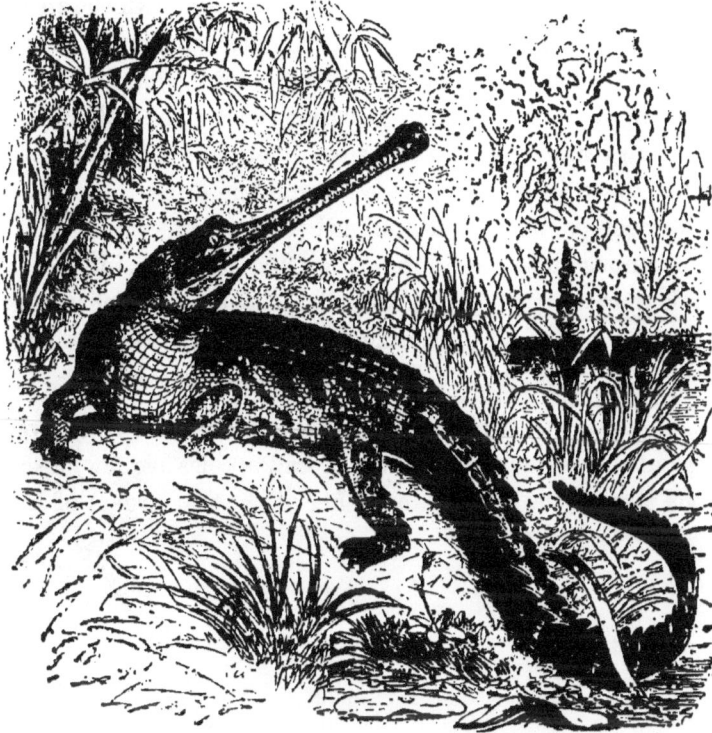

Fig. 83.—Gavialis gangeticus. (After Brehm.)

by its very long snout, the Crocodile of the Nile and the Alligator of the Mississippi. Most of them are fish-eating forms, but many of them lie in wait for the smaller Mammalia when they come to drink. Their aquatic habit is associated with a powerful compressed tail, and completely or incompletely webbed toes, but the legs are, nevertheless, strong enough to enable them to leave one pond and drag themselves to another. As in the other Reptiles, the horny epidermal covering is well developed and characteristic, but there exist also in the cutis bony

shields on the back and behind the breast bone, which encase
the creatures in an almost continuous coat of mail.   Important
differences from the Lizards exist in the endoskeleton; the
great elongation of the skull, *e.g.*, with the formidable teeth
lodged in sockets, the form of the nasal cavity, with the anterior
nostrils at the tip of the snout and the posterior far back in the
mouth, the intimate union of all the cranial and facial bones
with each other, etc.   Many peculiarities of the other organs
point to a greater specialisation than exists in the other Reptiles,
thus the brain is of a higher type, and the heart is subdivided
into four compartments, although there is still a certain mixture
of the venous and arterial blood immediately outside it.   The
order is oviparous like most Reptiles, the eggs, which are very
fragile from the small percentage of lime in the shell, being laid
in the sandy banks of the streams in which they live.

22. Although the Crocodilia hardly number more than twenty
species at the present day, yet the Fossil species are far more numerous than is the case in any other of the four orders of living

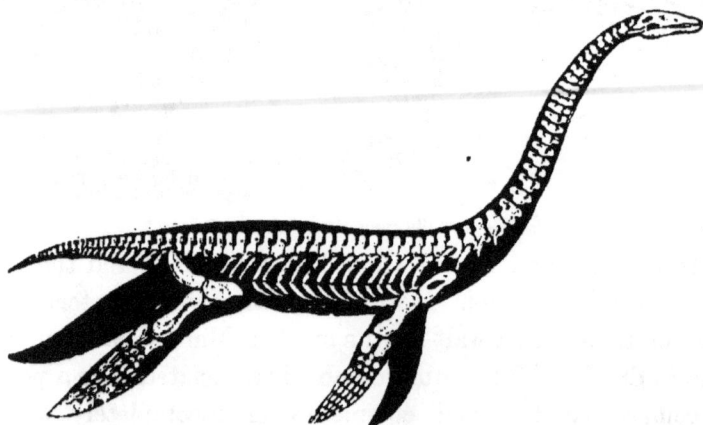

Fig. 84—Restoration of Plesiosaurus. ⅟₁₆.

Reptiles.   The earliest of them had amphicœlous vertebræ, and
a much more complete exoskeleton than the living Crocodiles,
probably for protection against the gigantic aquatic Reptiles

which inhabited the seas along with them. These arrange themselves under several orders of which the Sauropterygia (*Plesiosauria*) and the Ichthyoptergia (*Ichthyosauria*) are the best known. To the former (Fig. 84) belonged huge forms from 10-50 feet in length, with flippers something like a seal's, and an extremely long swan-like neck, which must have allowed great freedom of movement to the head with its formidable teeth. To the latter belonged short-necked forms resembling in shape the whales of the present day, but provided with a long and powerfully toothed snout. (Fig. 85).

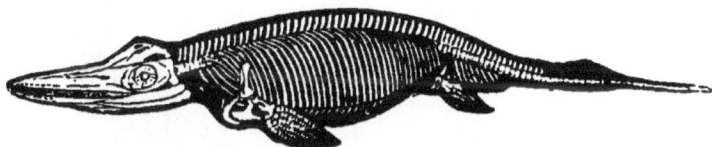

Fig. 85—Restoration of Ichthyosaurus. $\frac{1}{10}$.

25. Among the fossil orders are likewise forms which attained a huge size, whose limbs, more lizard-like in form (**Sauropoda**), attest to a terrestrial or amphibious life, but whose teeth indicate that they were herbivorous animals feeding either on aquatic or marsh plants or on the forest vegetation. (Fig. 86.) Some

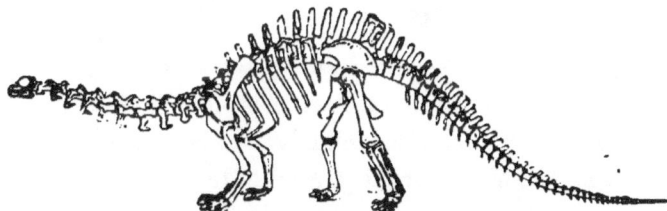

Fig. 86—Restoration of Brontosaurus. $\frac{1}{100}$.
(After Marsh).

of them (*Atlantosaurus*) measured 100 feet in length by 30 in height, the locomotion of such enormous masses being only rendered possible by the fact that the skeleton was extremely light, the bones being filled with air. The Sauropoda were without the protection of an exoskeleton, whereas Stego

saurus had bony shields in the skin and projecting horns from
the back which must have afforded a very complete defensive
armour.   This genus also is interesting from the fact that the
fore legs were shorter than the hind, and consequently that the
latter along with the tail supported the weight of the body.   A
transition is thus afforded to the **Ornithopoda** or bird-footed
Dinosaurs, a remarkable group, of which the best known is the
genus **Iguanodon** of the Cretaceous period.   (Fig. 87.)   Recent

Fig. 87—Skeleton of Iguanodon in the Brussels Museum. $\frac{1}{70}$.

discoveries in Belgium have disclosed complete skeletons of this
reptile, which is characterised chiefly by the strong bird-like
three-toed legs, the short fore legs used only for prehension, the
lizard-like tail and the compressed body.   The foot prints of this
Iguanodon, which are also preserved, show that its gait was
erect, and this is confirmed by examination of the sacral region
of the vertebral column, which is formed of five or six united
vertebræ, evidently for the purpose of transferring the weight of
the body to the hind legs.   The Iguanodon was herbivorous, but
there were carnivorous Dinosauria likewise, some like Compsognathus (Fig. 88) of such small dimensions that they hardly

deserve the ordinal name, others, like Megalosaurus, rivalling the largest herbivorous forms in size. Many of these carnivorous forms present features in their limbs and teeth which remind us of the carnivorous mammals, but the Compsognathus had an erect gait like the Iguanodon.

26. In many respects the Iguanodon and its allies resemble the Ostriches, and, indeed, as we shall see there are fossil toothed birds which help to fill up the gap between them.

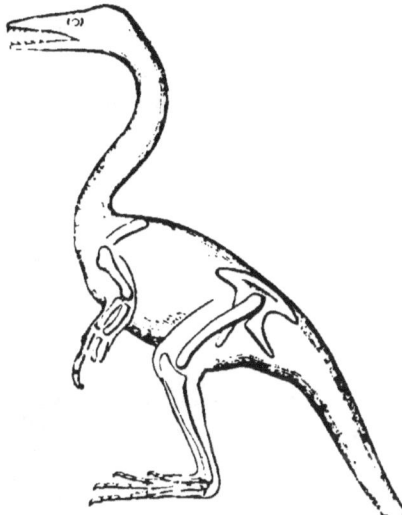

Fig. 88—Restoration o Compsognathus. ⅛.

So far the fossil reptiles we have considered have been either aquatic or terrestrial forms ; some of the latter indeed walked

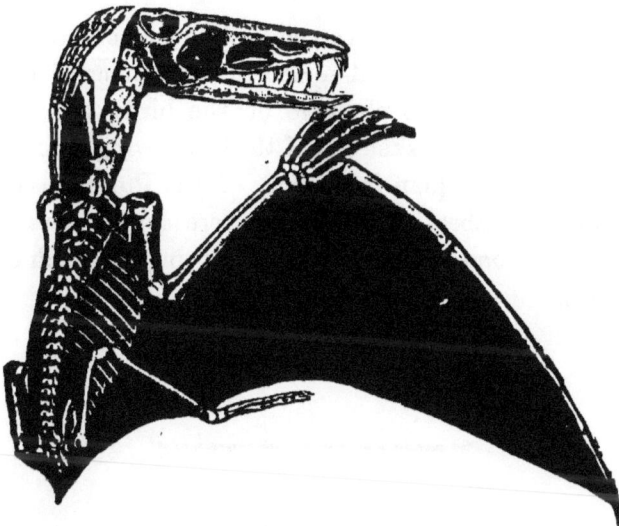

Fig. 89—Restoration of Pterodactyl. ⅛.

erect, their forelegs being used for prehensile purposes. We now come to certain forms which lived an aerial life, being pro- vided with organs of flight of a character peculiar to themselves· In this order, the **Pterosauria**, (Fig. 89) the limbs were approxi- mately of the same size, but the little finger of the anterior extremity was enormously long and strong compared with the others. It had four joints, and between it, the arms and the side of the body, a web of skin was stretched out, somewhat similar to the web between a bat's fingers. Some of them (Fig. 90)

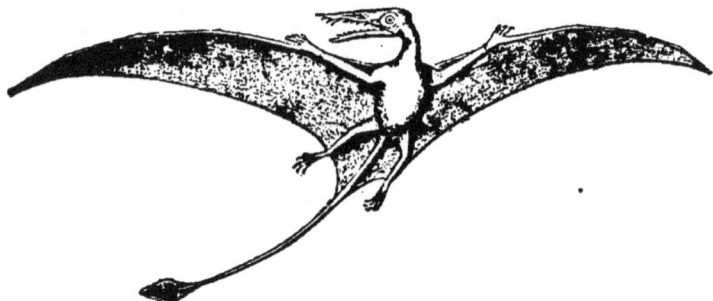

Fig. 90—Restoration of *Rhamphorhynchus phyllurus.* (Marsh). ⅓.

had a long tail terminating in a rudder-like membrane supported by the spines of the terminal vertebræ. But none of them had anything similar to the plumage of the birds, and we shall afterwards see that the skeleton of the bird's wing is constructed on quite a different plan from that of the Pterosaurian.· Al- though many members of this group were small, others attained a gigantic size, one found in Kansas having a stretch of some twenty feet.

CHAPTER V.

THE BIRDS.

1. From what we learned in last chapter it is neither the power of flight nor the supporting of the body by the hind limbs which constitutes a bird, for both these characteristics were present in certain fossil reptiles. It is the peculiar epidermal clothing of feathers, which we must regard as marking off the birds from the flying reptiles to which they are allied. Yet there is not so much difference between the reptilian scales and the avian feathers as might at first sight appear. If we watch the development of the feathers in a bird, we may see that they arise at first very much like scales, along regular tracts, and that they are simply thickenings of the epidermis over papillæ of the cutis. But as the feather is developed it is retracted into a follicle in the skin, and the epidermis gives rise to the feather proper, the dried-up cutis-papilla to the pith.

2. In most birds we distinguish two different elements in the plumage, the feathers proper or contour-feathers and quills, and the down-feathers; but in certain birds destitute of the power of flight down-feathers alone are present. Both kinds resemble each other in having a quill by which they are inserted into the feather-follicle, a shaft, and a vane which is composed of two rows of barbs, each provided with projecting barbules. If the barbules are so arranged that those on contiguous barbs interlock with each other, then we have a contour-feather, or if it is of special use in flight, is a quill-feather, but if the barbules are soft, and do not interlock, we have the down-feather. It is obvious that a creature only possessed of the latter cannot fly ; the interlocking of the barbules is neces-

9

sary to give to the feather the requisite strength to encounter
the resistance of the air.   The feathers, however, are not to be
merely regarded as organs of flight, but as a warm clothing for
the body, necessary to prevent the great loss of heat which
would otherwise attend the quick flight and rapid change of air
round their owners.

3. A great many technical terms are necessary in systematic
ornithological descriptions ; some of these may be studied from
Fig. 91.

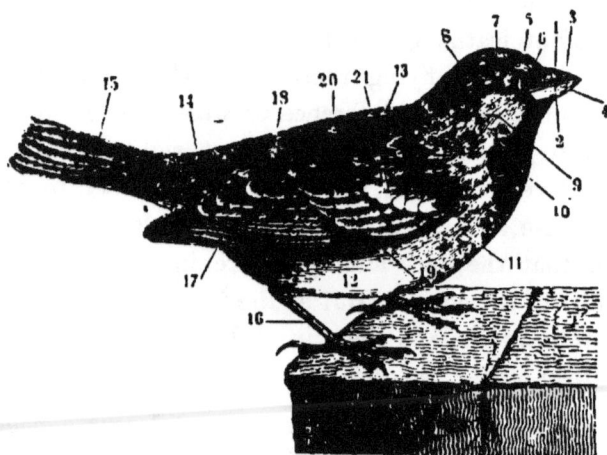

Fig. 91—To illustrate the topography and the plumage of the Sparrow.

(*Passer domesticus*).   (After Thomé).

1, Upper mandible with nostril ; 2, lower mandible ; 3, culmen ; 4, gonys; 5, lore in
front of the eye ; 6, forehead or frons; 7, crown or vertex; 8, occiput; 9, malar or cheek
region ; 10. throat; 11, breast ; 12, abdomen ; 13, back ; 14, rump with the upper tail
coverts ; 15, rectrices or tail feathers ; 16, tarsus ; 17, primaries ; 18, secondary and
tertiary remiges, or wing quills; 19, alula or bastard wing ; 20, greater ; 21, median and
lesser wing coverts.

4. Premising then that the plumage is to be regarded as
the essential characteristic of a bird, let us see, in continuation
of the subject of the close of the last chapter, how early
the remains of true birds are to be met with in the geological history of the earth.   Fortunately the feathers are
well known in the earliest fossil bird which has been found,

because the remains are preserved in a very fine textured stone of UpperJurassic, Agefound at Solenhofen in Germany, andusedfor lithographic purposes. The last found of the two specimens is very perfect, the carcase of the bird having been bedded in the fine sediment of the sea-shore, in such a position that all the parts are very plain. From these we recognise that the *Archæopteryx*, as it is called, is in many respects more like a reptile than a true bird, especially so in the fact of its tail being formed of a large number of distinct vertebræ, so that the ordinal name **Saururæ**, was formed for it on this account. (Fig. 92). No other bird-remains hitherto found show this peculiarity, so that Archæopteryx stands alone with its lizardlike tail, while all other birds, living and fossil, show some union of the caudal vertebræ. In

Fig. 92—Berlin specimen of Archæopteryx, ⅓, with leg from London specimen. UF. Tibia. MF. Tarso-metatarse. Z. The toes.

respect to its plumage and organs of flight, however, Archæopteryx is a true bird, and not a flying lizard. Two other fossil birds have been found in the Cretaceous rocks of the Western States which share with Archæopteryx another reptilian character, that of toothed jaws, but in other respects more closely resemble the birds of the present day Of these two genera, one, *Hesperornis*, appears to have been destitute of the power of flight, because the bones of the anterior extremity are much reduced and there is no keel upon the sternum, such as there is in all birds which have that power, while the other, *Ichthyornis*, had a wing fashioned in the same way as the great majority of our birds, necessitating a keel upon the sternum. Thus at a very early period of the history of the birds, the distinction which we draw between those with a raft-like and those with a keeled sternum (the **Ratitæ** and the **Carinatæ**) was already established. But this distinction is of a comparatively unimportant character, because it would appear that at various periods of the world's history, members of different families of birds have lost the power of flight (and along with this, to a greater or less degree, the keel on the sternum,) by taking exclusively to some other method of locomotion, such as swimming or running. Nevertheless, when we realize how the structure of a bird is connected with its power of flight, we shall more easily understand how the absence of that power may produce a superficial resemblance.

5. If we study, then, the structure of any typical carinate bird, we shall soon learn that apart from the plumage there are many other features which are evidently adapted to its mode of locomotion and habits of life, and that the whole structure of the body is, indeed, modified in connection therewith. Although the common domestic fowl belongs to a family of poor fliers, yet a knowledge of its structure forms a key to that of all the Carinatæ. Numerous races are known but all belong to a single species, *Gallus domesticus*, the nearest wild ally of which is the *Gallus bankiva* of India. With the pheasants, pea-fowl and guinea-fowl they form the family **Phasianidæ** of the order **Gallinacei**, all of the birds included under which (prairie-fowl, partridge, turkey, etc.,) are indifferent fliers, seeking their food on the ground, partly by scraping, for which purpose the feet are provided with strong claws.

6. Let us now proceed to examine the skeleton of a fowl with the object of seeing in what respects it differs from that of the fossil birds and reptiles   The vertebral column presents the five

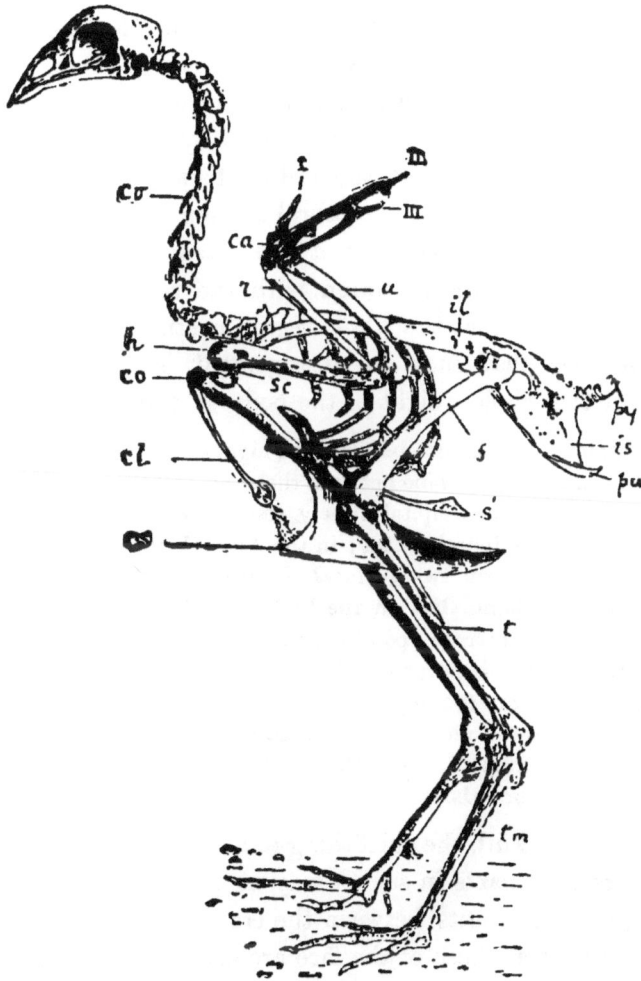

Fig. 93.—Skeleton of Fowl.

cv. cervical vertebræ ; h. humerus ; r, u. radius and ulna ; ca. carpus ; I. pollex ; II index ; III. middle finger ; co. coracoid: sc. scapula, cl. clavicle ; cs. crest of sternum; s'. one of the processes of the body of the sternum ; il. ilium ; .s. ischium ; pu pubis; py. pygostyle ; f. femur ; t. tibia and fibula ; tm, tarso-metatarse.

regions present in all the higher Vertebrates ; of these, the cervical contains thirteen vertebræ, and it alone retains any freedom of movement between the individual bones, for the vertebræ of the trunk and the greater part of the tail are for the most part so united by bone that no such movement is possible. The last few vertebræ are, however, movable, and the terminal one, the so-called *pygostyle* (which is really formed by the fusion of several centra) serves to support the tail-feathers. The chief characteristic, then, of the bird's vertebral column is that the iliac bones, instead of being united with a few sacral vertebræ as in the reptiles, have acquired a union with the vertebræ in front and behind these, so that the whole region is stiffened into one mass. It is obvious that such an arrangement is well calculated to transmit the whole weight of the body to the hind limbs.

A conspicuous difference between the higher and the lower Vetebrates is in the nature of the union between the skull and the vertebral column. In the fish the basi-occipital closely resembles and forms a part of a series with the vertebral centra ; in the Amphibian, on the other hand, the ex-occipitals form two surfaces of contact (**occipital condyles**) with the vertebral column, but in the higher Vertebrates greater freedom of movement is conferred upon the head by the specialisation of the first two vertebræ for this purpose. The anterior or **atlas** possesses surfaces for receiving the one (Sauropsida) or two (Mammalia) occipital condyles, and it itself rotates on a pivot (the **odontoid process**—which is really part of the centrum of the Atlas—) attached to the centrum of the second vertebra or **axis.**

As for the skull, the chief thing to notice is the early fusion of its component parts in such a way as to make it impossible to distinguish the limits between the cranial bones, although the sutures between those of the face are easily seen. The premaxillæ are of large size, and chiefly support the upper beak, (which is movable 10 a certain extent on the skull behind), while the maxillæ are insignificant and are united to the outer surface of the movable quadrate by a slender bar formed of the jugal and quadrato-jugal. The quadrate moves partly on the skull and partly on the pterygo-palatine bar, and it supports the mandible, which in the adult is formed of a single piece inclosed in the lower horny beak.

It is to be expected that the possession of the power of flight should be associated with alteration of those parts of the skeleton concerned, so that in addition to the rigidity of the thorax conferred by the immovability of the dorsal vertebræ, it is not surprising to find other conditions adapted to offer a solid basis of resistance to the stroke of the wing, and protection to the delicate parts contained within the thorax. So the true ribs are not only bound to each other by **uncinate** processes, but their sternal ends, instead of being cartilaginous, are bony. Again, the sternum does not only afford protection to the thoracic organs by its great size, but by its keel offers a large surface for the attachment of the muscles of flight. Although the scapula is small, the ventral parts of the shoulder-girdle are both strong and connected with the sternum; especially is this true of the coracoid, through which the strain of the wing stroke is chiefly transmitted to the trunk. The chief peculiarity of the humerus is its strong crest for the insertion of the muscles of flight, while the ulna differs from the radius in its strong curvature, the convexity of which is roughened by the attachments of the secondaries. It is the wrist and hand that are most peculiar, however, for we see only two proximal carpals, the distal carpals being fused with the three metacarpals into one perforated bone, while the three fingers are independent. The second finger is the most important, the third being rudimentary, while the first, which supports the spurious wing when it is present, and like the second is sometimes provided with a claw, is also short.

Reference was made to the singular method of union of the pelvis to the trunk; the other parts of the pelvic arch are chiefly remarkable for their backward direction and for their not meeting in a symphysis, except in some of the more reptile-like Ratitæ. Another peculiarity of the hinder extremity is that only the proximal end of the fibula is present, and that the ankle joint, as in many reptiles, is situated between the proximal and the distal rows of tarsal bones, the former of which becomes fused with the tibia, into a tibio-tarsus, while the latter and the metacarpals of the second, third and fourth toes become fused into one tarso-metatarse. When the first toe is present, its metatarsal is generally rudimentary, but it has two joints, while the second toe has three, the third four, and the fourth five. The spur of the cock is simply a bony excrescence attached to the metatarsus.

7. It is to be expected that the greatest peculiarities of the muscular system of the birds should be connected with their mode of locomotion. From what has been said above, it will be gathered that the great muscles of flight take their origin from the sternum, and thus the centre of gravity of the body is shifted towards the most favourable position for flight. Not only the depressor of the wing, but also its elevator muscle arise from the sternum, the necessary change in the direction of the latter being acquired by its tendon passing through a pulley, at the junction of the three bones of the shoulder-girdle, to its insertion in the upper surface of the humerus. In reptiles and mammals where the full number of fingers is present, and the joints of these are freely movable upon each other and on the wrist bones, muscles are necessary for carrying out these movements, but the consolidation in the region of the hand of the bird dispenses with the necessity for these and therefore the chief muscles of the fore limbs are in its proximal end near the body. The same is true of the hinder extremity, tendons only being continued into the distal end to carry out the movements of the toes. Thus the great muscles of the limbs are likewise situated in a favourable position of the body for flight. Cutaneous muscles, chiefly inserted into the feathers

and destined for shaking the plumage, attain a greater development in the birds than they do in the lower forms.

8. A great advance is to be seen in the brain of a bird as
compared with that of any reptile, for not only is the cerebrum much larger, but the cerebellum is so also, with the result
that the optic lobes are thrust aside right and left towards the
base of the brain. The surface of the cerebrum is smooth, but
that of the cerebellum is much folded so that the white matter
is arranged in a tree-like fashion in its interior. Of the senses,
sigth is decidedly the most acute; the birds of prey especially are
gifted with extraordinary powers of vision, and in association
therewith the bulb of the eye has a very different shape from
the globular one present in other Vertebrates. Its principal
axis is much elongated, the posterior part of the bulb being a segment of a sphere, while the anterior is drawn out in a tubular
fashion. As in some of the reptiles, the sclerotic coat has a
circlet of bony plates formed in it. Hearing is also more acute
than in the reptiles; the tympanic membrane is situated at the
bottom of a short external auditory passage (surrounded by
special " auricular" feathers) and the Eustachian tubes converge to a common aperture in the palate.

9. Although organs both of touch and taste are present in
the mouth-cavity of the bird, yet the tongue is generally clad
to a great extent with horn, which varies in shape in different
species. The most constant peculiarity of the œsophagus is the
presence of a crop, which may be a projection from one side, as
in the fowl (Fig. 94), or from both, as in the pigeon. The
stomach is divided into a smaller glandular cardiac end, the
proventriculus, and a larger muscular pyloric end, the gizzard.
In the latter the muscular coat is very thick in two places, and
the epithelial lining is converted into horny pads, which serve
for the grinding of the food in the granivorous birds. As a
compensation for small salivary glands, the pancreas is large,
and the length of the intestinal surface is increased by two

cœca, which open into the anterior end of the large intestine.

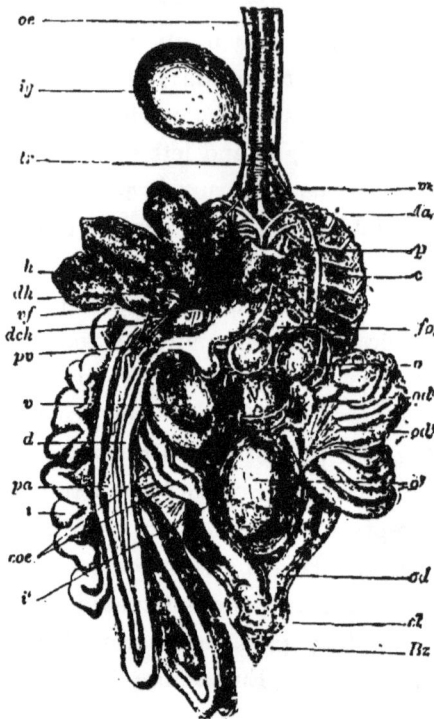

10. Certainly the most remarkable feature about the respiratory system of birds is the development of air-sacs, which receive their air from the lungs, and are situated partly among the viscera of the body-cavity, partly between the muscles, and underneath the skin of the body, and finally within the bones, displacing the marrow in these. The function of these air-sacs is not confined to supplementing the size of the lungs, (although a certain interchange of gases between the blood and the contents of the air-sacs must take place), but they likewise serve to render the body specifically lighter

Fig. 94—Viscera of the Fowl. (after Brandt).

oe, œsophagus; ig, crop; tr, trachea; m, muscle; la, syrinx; p, lung; c, heart; h, liver; dh, hepatic duct; vf, gall-bladder; dch, bile-duct; pv, proventriculus; sp, spleen; v, gizzard; d, duodenum; pa, pancreas; i, small intestine; coe, its cœca; fo, egg-follicle burst; o, eggs; od, oviduct partly slit open containing a mature egg, o'; cl, cloaca; Bz, Bursa Fabricii.

(especially as the bodily temperature is high), and thus more adapted for flight. As the bird's locomotion involves much muscular exertion, both the respiratory and circulatory systems are more perfect than in the reptiles, the presence of four chambers in the heart, allowing the complete separation of the blood which has been returned for aeration, from that which is sent out by the heart to the system.

11. In the lower Vertebrates the voice is little developed,

but in the birds, especially in the song-birds and parrots, it is not only of considerable range, but also capable of modulation. The larynx, which is the organ of voice in the other Vertebrates, is here in the background, for the notes of the song-bird are produced lower down in the wind-pipe, at the point where it divides into the two bronchi. Here the "syrinx" is situated, in the formation of which, both membranes capable of vibration, and resonant dilatations capable of reinforcing the sounds produced by these, take part.

12. The kidneys in the birds are not so elongated as in the reptiles, and are moulded into the large and complex sacrum.

Only one oviduct, the left, is present ; the number of eggs laid is very different in different species, but approximately constant in the same species. The size is not always directly proportionate to the size of the bird, for the chicks escape from the egg (by the agency of a temporary tooth on the upper beak) at very different periods of their development. Of the various orders of Carinate birds, those which are the more primitive escape from the egg in a condition to fend for themselves (**aves precoces**), while the young of the higher orders required to be looked after by the parents for some time after they are hatched, their escape taking place at a much less developed phase (**aves altrices**).

13. The egg of the fowl owes its large size chiefly to the food-yolk, which is associated with the germinal yolk (I, 66). The yolk is, nevertheless, a single cell bounded by a wall, the vitelline membrane. After bursting through its capsule in the ovary (Fig. 94), it escapes into the cœlom, and is received by the open mouth of the oviduct, the walls of which are provided with glands, which secrete the albumen or white, and with muscular layers, which propel it in a spiral direction (involving the formation of the ropy parts of the white) towards the lower end of the tube, where the shell-glands secrete the shell. When the egg is laid, it has already undergone some of the stages of segmentation, the white patch upon the surface being formed of a layer of cells (the blastoderm), destined to grow into the body of the chick. The process of incubation requires twenty-one days, and it can be carried out

artificially, if fresh air and moisture, as well as the proper temperature —104° F.—be afforded to the eggs.

14. Having examined the structure of the fowl as a convenient carinate type, let us now see in what respects Archæopteryx and the Ratitæ differ from it. As far as the plumage is concerned, Archæopteryx approaches the Fowl more closely than do the Ratitæ, for in the latter the feathers are mere downs, while in the former quill-feathers were present, and probably also fine contour-feathers, although the impressions of these have not been preserved. The quill-feathers were attached to the ulnar side of the hand and fore-arm, round the neck, to the leg as far as the tarsal joint, and in a single series along each side of the long tail. That is really the most important peculiarity of the fossil, for instead of the short tail of the Fowl, there were twenty independent vertebræ, each with a quill-feather attached right and left to it. The trunk region, likewise, shows less of the concrescence so marked in the carinate bird, for the the vertebræ are all amphicœlous, and only a few of them are united into a sacrum; furthermore, the ribs are decidedly reptile-like in their arrangement. In place of the horny sheath of the bird's bill, Archæopteryx was furnished with numerous little conical teeth, probably lodged in sockets; in other respects the skull was bird-like. It is argued from the absence of a crest on the humerus that Archæopteryx was a poor flier, (the sternum has not been found, so we lack the evidence which would have been forthcoming from it), but the hand was formed of three fingers with independent metacarpals and stout claws, so it is likely that the anterior extremity must have been of great service in climbing, the plumage serving, perhaps, more as a parachute than for true flight. In the structure of the hinder extremity there is no great difference from that of a bird.

15. The Ratitæ or Cursorial Birds are unquestionably much closer to the Carinate Birds than Archæopteryx is; indeed many

zoologists regard them as degenerate forms of carinate birds, which have in the course of ages lost the power of flight, while others, looking at their structure and their geographical distribution, think that they are a more primitive group than the Carinatæ with more affinity to the reptiles, and that they never possessed the power of flight. The plumage in this group never

Fig. 95.—Skeleton of the Moa (*Dinornis*).

has the character of contour or quill-feathers, the plumes of the ostrich being nothing but gigantic downs. Instead of the bones of the head uniting early with each other, the sutures are quite evident, and the cervical ribs are for a long time movable, instead of being coalesced with the vertebræ. Again, either the pubic or ischiac bones or both may form a symphysis as they do in most living reptiles ; and, at least, there is a greater resemblance to the Dinosaur pelvis than there is in the carinate birds. Other anatomical features of the Ratitæ are adaptive ; the functionless nature of the fore-limbs is associated with the reduction in size of their bones and of the clavicles, while the adaptation of the hind limbs to rapid locomotion leads to a loss of one or two of the four toes.

There is great structural difference between the families of Ratitæ, and they are also marked off geographically from each other. New Zealand has the Kiwis (*Apteryx*), small forms of about the size of a turkey with a very rudimentary anterior extremity and four toes, of which the hinder one is strongly clawed. Allied to it are the remains of various giant birds (*Dinornithidæ*), recently extinct and found for the most part in New Zealand (Fig. 95). These Moas, as they are named by the Maoris, stood ten feet from the ground, and their eggs were of very large size. Allied to them is a similar form from Madagascar (*Æpyornis*), believed to be the Roc of Eastern Fables ; the skeleton of this genus is not well known, but eggs have been found of enormous size, which hold as much as two-and-a-half gallons, and have been estimated to be equivalent in contents to twelve dozen hen's eggs. In the rest of the Australian region two other genera, the Cassowaries (*Casuarius*), and the Emus (*Dromæus*), are found, while in South America, are the three-toed Ostriches (*Rhea*), and in the deserts of Africa and Western Asia, the two-toed Ostrich (*Struthio camelus*).

Like the Carinatæ the Ratitæ have no teeth in the jaws, which are simply clothed with a horny beak, but the genus *Hesperornis*, which as far as its sternum is concerned is one of the Ratitæ, had only a horny beak on its premaxillæ, while the maxillæ and the mandibles had teeth fixed in a continuous groove. Besides the teeth, there were numerous other characters which give it an intermediate position between the Dinosaurs and the Ratitæ. The anterior extremity is represented by the humerus alone, but the whole skeleton gives the impression of a large diving bird like a Grebe, living on fish, and swimming by m..ans of the powerful feet. On the other hand, the genus *Icththyornis* was truly carinate, its anterior extremity being like that of an ordinary bird, but the rest of the skeleton presenting primitive features indicating reptilian affinities, such as teeth arranged as in Hesperornis (except that they were in sockets), and amphicœlous vertebræ as in Archæopteryx. Icththyornis, therefore, occupies a middle position between Archæopteryx and the Carinatæ.

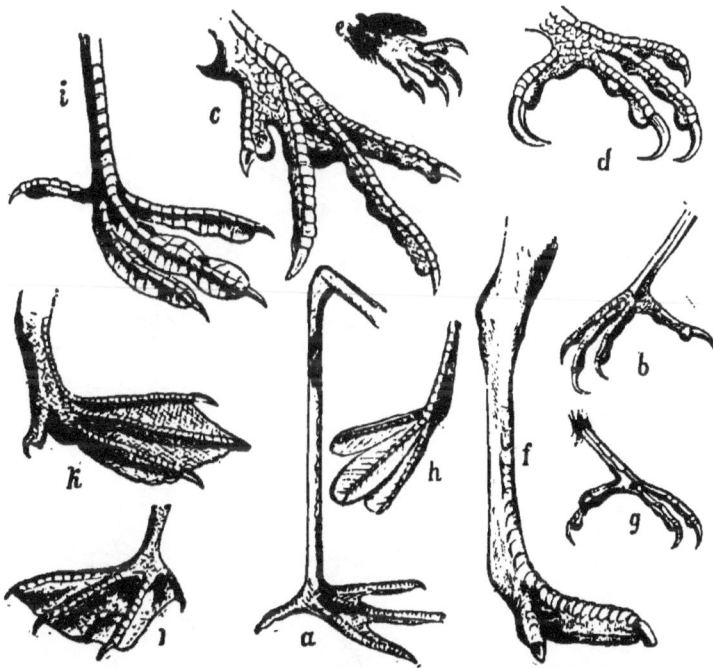

Fig. 96.—Feet of various Avian genera.

a. wading type, *Ciconia*; b. perching, *Turdus*; c. rasorial, *Phasianus*; d. raptorial, *Falco*; e. adherent, *Cypselus*; b. cursorial, *Struthio*; g. scansorial, *Picus*; h. lobate, *Podiceps*; i. lobate and scolloped, *Fulica*; k. palmate, *Anas*; l. totipalmate, *Phaethon*.

16. When we come to the classification of the Carinate birds we meet with great difficulties; for although we recognise that there are certain orders which are lower than the others, yet the adaptation to an aerial life has impressed a certain uniformity

upon all, concealing such structural characters as might be relied upon for making a natural classification, and causing the ornithologist to depend frequently on characters which are in relation to the food or the manner of life (Figs. 96, 97). The

Fig. 97.—Outlines of bills of various genera.

L. *Leptoptilus*, marabii; P. *Passer*, sparrow; Ca. *Cancroma*, Boatbill; D. *Docimastes*, Swordbill; Pl. *Platalea*, spoonbill; Pe. *Pelecanus*, pelican; T. *Turdus*, hrush; Re. *Recurvirostra*, avocet Ph. *Phœnicopterus*, Flamingo; Ry. *Rhynchops*, Skimmer; A. *Anastomus*, stork; B. *Balœniceps*, shoebill; S. *Sarcorhamphus*, condor; Co. *Columba*, pigeon; My. *Mycteria*, stork; Me. *Mergus*, Merganser; 1. Ibis.

difficulties of classification are chiefly met with among the higher orders, to which not only by far the greatest number of the species belong, but which exhibit far less important differences between each other than do the members of the lower orders.

17. The following arrangement of the orders of Carinatæ is that generally employed ; although there may be doubt as to the affinities of some of the groups, there is none that the swimming birds occupy the lowest place and the song birds the highest.   Of the former the **Pygopodes,** or Divers, are marked by the far back position of the legs required by an erect or semi-erect attitude, by the shortness or rudimentary character of the wings, by the complete or incomplete webbing of the three toes, and by their powers of swimming and diving.   The most remarkable genus is the Penguin (*Aptenodytes*) of Southern Seas, (Fig. 98) in which the power of flight has been lost, and the wings are converted into flippers covered with scale-like feathers.   Another interesting form, in which the wings, though feathered, were extremely short and incapable of flight, is the great Auk (*Plautus impennis* (Fig. 99), which was common in

Fig. 98.—Penguin (*Aptenodytes*)
(after Brehm.)

the Arctic Seas at the beginning of this century, but is now thought to be quite extinct. Allied to it are the Puffins (*Fratercula*) with their singularly shaped and brilliantly coloured bills, and the Sea-pigeons (*Cepphus grylle*)

1v

the most elegantly formed of the group.   More familiar than these marine forms are the Loons (*Urinator*) and the Grebes (*Podicipidæ*), the latter with the toes merely fringed, the former with the toes entirely webbed.

18. In contrast with these forms are the long-winged swimmers **(Longipennes)**, in which the length of the wing is due to the length of the arm-bones, not of the hand.   They are excellent fliers, and sweep down upon the sea and inland lakes for the aquatic animals (chiefly fish) on which they live. The three front toes are webbed, the first, free and often rudimentary.   The gulls (*Larus*), Terns (*Sterna*), and the Jægers (*Stercorarius*) all belong to this group.   Nearly allied to it are the marine petrels, to which small forms like the stormy petrel or Mother Carey's chicken (*Procellaria pelagica*) belong, and the giant albatross, *Diomedea exulans*, with a wing-spread of 15 feet.   The nostrils of the latter, however, are tubular, not mere fissures as in the gulls, etc. ; they are, therefore, often regarded as a distinct order **(Turbinares)**.

19. The **Steganopodes** have received their name from the complete webbing of the toes, the first toe being turned forward and united by a mem-

Fig. 99.—Great Auk. *Plautus impennis.*
(after Brehm.)

brane to the second.   They are all fish-eating birds, but embrace such different forms as the tropic and frigate birds (*Phaethon* and *Tachypetes*), the darters (*Plotus*), gannets (*Sula*), and, more familiar in-land, the cormorants (*Phalacrocorax*), and pelicans (*Pelecanus*).   The singular mandibular pouch of the last genus marks it from the others.

20. Unlike the above, the Ducks and their allies **(Anseres)** have the

hinder toe free ; the others are webbed, and the beak is covered with a soft skin in which there are numerous tactile corpuscles, while the gape is provided with horny lamellæ (hence *Lamellirostres*), which serve for straining the muddy water in which they seek their food. The least duck-like forms are the Mergansers, which have a serrated bill and dive for fish. A very large number of species of wild ducks are known, from one of which, the Mallard (*Anas boschas*), the domestic duck is derived. To the same genus belong the Teal and Widgeon, but the Shoveller (*Spatula*), Pin-tail (*Dafila*), Wood-duck (*Aix*) and Red-head and Canvas-back (*Aythya*) are sufficiently different to be separated under distinct genera. The same is true of the Buffle-head (*Charitoneta*), Harlequin (*Histrionicus*), the various species of Eiders (*Somateria*) and Scoters, and the Ruddy Duck (*Erismatura*). The domestic goose is derived from the European *Anser cinereus*, which genus is represented in America by the white-fronted goose. Various allied forms, like the Canada goose and Barnacle goose, are ranged under the genus *Branta*. To the same order belong the Swans (*Cygnidæ*), and allied to it are the Flamingoes (*Phœnicopterus*) with their singular bent bills, very long legs and brilliant plumage.

21. In the next order (**Herodiones**),* we have various genera which like the Flamingoes have very long legs, the tibia and tarsus being much elongated, but they differ from them in the structure of the bill, and also from the following order in the same respect. The bill has no cere or fleshy part at the root as in the other waders, but it is very differently shaped in the different genera, *e.g.*, in the Spoonbills (*Platalea*) it is flattened and spatulate at the tip, in the Ibises compressed and arched downwards, in the Storks (*Ciconia*) much thicker than in the Herons (*Ardeidæ*), from which the order derives its name. This family embraces the Herons (*Ardea*), Bitterns (*Botaurus*), the Night Herons (*Nycticorax*) ; the rest of the waders are subdivided into marsh-birds and shore-birds. To the former (**Paludicolæ**) belong the Cranes (*Gruidæ*), and Rails (*Rallidæ*) including the Gallinules and the Coots, while the latter (**Limicolæ**) embrace the Avocets (*Recurvirostra*), Snipes (*Gallinago*,) Woodcock (*Philohela*), Sand-Pipers (*Tringa* and *Totanus*), Curlews (*Numenius*), and Plovers (*Charadrius* and *Ægialitis*), etc.

22. In contrast to the long-legged Waders we now come to the **Galli-nacei** the legs of which are short, stout, and adapted for scraping. The Pheasant family (*Phasianidæ*), to which the domestic fowl belongs, is only represented in America by the wild turkey, *Meleagris gallopavo*, (the probable stem-form of the domestic turkey), but it is abundantly repre

sented in the Old World, and especially in India by the Pheasant (*Phasi-anus colchicus*), Peafowl (*Pavo*), Guineafowl (*Numida*), Argus Pheasant (*Argus*), etc. On the other hand, the Grouse family is as character-istically American as Old World, for we have Ruffed Grouse (*Bonasa*) Prairie-hen (*Tympanuchus*), Ptarmigan (*Lagopus*), and other forms. Two aberrant families are associated with the Gallinæ which exhibit primitive characteristics in two different directions,—the **Megapodidæ** of Australia, which do not hatch their eggs, but lay them in heaps, to be incubated by the heat evolved from decomposing vegetable matter mixed with them and the Tinamus of South America, which in the structure of the skull remind us of the Ratitæ.

23. The Pigeons or **Columbæ** are better adapted for flight than the, foregoing order, the feet are more delicate, the bill has a soft cere, and the young are looked after by the parents on their escape from the egg. In this region we have merely the passenger pigeon (*Ectopistes*) and the mourning dove (*Zenaidura*) ; but the pigeons form a very large group, especially developed in the Australian region, where the ground-pigeons (*Goura*) and fruit-pigeons (*Carpophaga*) are abundant. Our numerous domestic races and varieties are all derived from the Mediterranean Rock-pigeon (*Columba livia*). Now-extinct forms allied to the pigeon were the Dodo of Mauritius (*Didus ineptus*) and the Solitaire of Rodri-guez (*Pezophaps*), their extinction being attributable to their rudimentary wings.

24. All the birds of prey (**Raptores**) agree in the possession of strong curved claws and bill, the upper beak projecting like a hook beyond the lower, and with a cere surrounding the nostrils. The tarsus may be scutellate or partly feathered. It is naked and extremely long in the Secretary Bird, *Gypogeranus*, a crane-like form from the African steppes, which chiefly hunts Reptiles, but it is comparatively short in the other genera, especially so in the Owls, where not only the tarsus is feathered, but also the foot. The whole plumage in the Owls is of such a character as to permit the noiseless flight so helpful to them as nocturnal birds. Their habit of concealing themselves during the day in trees, rocks, etc., is assisted by the climbing foot. Some of the species like the Barn-Owl (*Strix*), and Saw-whet Owl (*Nyctale*) have complete radiating disks of feathers round the eyes, others like the Horned Owls (*Bubo*), only horns. The tiny burrowing Owls of the Western States (*Speotyto*) have the singular habit of nesting at the ends of long burrows.

The above-mentioned families include extreme types of the Raptores, between which are the vultures, eagles, falcons, etc., more closely related to each other. Among the vultures those of the New World (*Cathartidæ*) resemble those of the Old World (*Vulturidæ*) in the absence of feathers about the head, but differ in the structure of the bill, and in their habit of feeding on carrion. The Falconidæ, on the other hand, have a feathered head, shorter bill, and include the various buzzards, eagles, hawks, falcons and the ospreys.

25. A large series of very different forms used to be associated as the "scansorial birds" on account of their possession of "climbing" feet, but it is now recognized that they ought to be grouped under several orders. One of the most singular of them is that of the **Psittaci** or parrots, which are marked by the upper bill being shorter than it is high, strongly curved, movably articulated to the skull, and with a cere surrounding the nostrils. The lower bill, on the other hand, is short and truncated, and the tongue fleshy. This is essentially a tropical group, chiefly developed in South America and the Australian region. It embraces the cockatoo, macaws, parrots, and ground-parrots. Associated with the cuckoo under the ordinal name **Coccyges**, are a number of forms with the foot more or less adapted for climbing, and with a long bill, but forming, on the whole, a somewhat heterogeneous group. The Toucans of S. America with their gigantic bills, the Rhinoceros birds, with the singular horny process on theirs, the Cuckoos, Motmots, Kingfishers and Hoopoos belong to this order. The **Pici** are a more homogeneous order, embracing the Woodpeckers and Wrynecks ; the toes which are turned forward are connected at the base, and the bill is sharp and chisel-like. Finally, the **Macrochires** receive their ordinal name from the length of the hand, which is longer than the fore-arm, and that longer than the humerus. They are good fliers, and embrace the Goatsuckers, Swifts and Humming-birds.

26. More than half of the species of Birds belong to the last order, **Passeres**, in which the bill is differently shaped but always without a cere ; in accordance with their "perching" habits the hinder toe is longer and stronger than the second toe ; both the outer toes are connected at the base. They fall into two sub-orders, the *Clamatores*, which embrace the forms destitute of a syrinx such as the Flycatchers, and the *Oscines* or singing-birds, which include the Larks, Crows, Jays and Magpies, the Blackbirds, Orioles, Finches, Sparrows, Tanagers, Swallows, Shrikes, Creepers, Warblers, Wagtails, Wrens, Nuthatches, Tits, Thrushes and Bluebirds.

CHAPTER VI

The Mammals.

1. Among the classes of Vertebrates already studied we have observed much difference as to the care taken by the mother of her young. In the Fishes, for example, large numbers of eggs are produced, and for the most part left to their fate, only a certain number of the fry reaching maturity. But there are examples of nest-building forms among them, and of parents which defend and protect their young, while there are others in which the young are retained within the body of the mother until they are able to look after themselves. (I, 94.) Similar differences are to be met with in the Amphibia and Reptilia as to this point, and we have also seen that some Birds make very little provision for the safe hatching of their eggs, while in the case of others, the eggs are incubated and protected until the young escape, either able to fend for themselves or in a condition to require further protection and feeding from the mother. As an example of special adaptation to the care of the young may be mentioned the Pigeons, in which the glands of the crop secrete a milky fluid during the time of incubation, which is used for the nutrition of the squabs.

2. We meet with an analogous condition of affairs in the Mammals, where certain glands of the skin are specially adapted to furnish milk for the nourishment of the young. Even in this group, however, we see great differences as to the condition in which the young are brought forth, for puppies are born blind and helpless, while the young of the Herbivora are able to run about shortly after birth. If such differences are to be

observed in familiar animals, still greater differences characterise two orders of the Mammalia which are almost confined to the Australian Region—the Monotremes and the Marsupials. In the first of these groups eggs are laid, large in size, containing much food-yolk, and surrounded by a shell which has a certain amount of lime in its composition. After a short period of incubation the young escape from the shell in a very helpless condition, and are now dependent on the milk which exudes from the cutaneous glands of the mother. In one of the two genera which belongs to this order, the milk-glands open into a pouch, big enough to receive the head of the young, but in the other no such provision is present. The Marsupials, the second of these orders, however, in which the young are also born in a very helpless condition (the giant Kangaroo, as tall as a man, brings forth young of about the size of a newly-born rat), have a provision for sheltering them in the shape of the pouch, a fold of the skin on the ventral surface of the abdomen, which supports and protects the young while they are being fed by the milk-glands of the mother, and to which they resort in danger even after they are able to run about. In this respect, then, we are able to recognize three great groups of Mammalia, (1.) the oviparous forms; (2.) those which bring forth their young alive, but in such a helpless condition that they must be carried for a long time in the mother's pouch ; and (3.) the higher Mammals, in which the development of the young advances to a much higher degree within the body of the mother. These groups or sub-classes have been called **Prototheria, Metatheria,** and **Eutheria,** respectively, and while each of the first sub-classes contains only a single order of Mammalia, the third embraces all the remaining orders, and contains all the most familiar and conspicuous quadrupeds. It is not to be supposed that this primary subdivision into sub-classes depends alone on the characters mentioned; there are certain other anatomical differences in the skeletal and other systems, which, however, it will be easier to understand after we have glanced at the structure of one of the Eutheria, and compared it with that of the foregoing classes.

3. Any familiar mammal will serve our purpose equally well, but the Cat (*Felis domestica*) is perhaps as accessible an example of the group as we can study. It belongs to the order of the Carnivores or Beasts of Prey (**Carnivora**), and is accordingly furnished with claws and teeth which are adapted to its mode of life, and is, indeed, one of the most highly specialised of the order in this respect, so that we must not expect to find it a primitive example of Mammalian structure.

4. With few exceptions, the mammals are clad with a coat of hair, which like the plumage of the birds enables them to preserve their high bodily temperature. The exceptions to this rule are certain aquatic forms where a subcutaneous accumlation of fat, the blubber, furnishes the necessary non-conducting envelope, and certain terrestrial forms, where the epidermis is either extraordinarily thick as in the Pachyderms, or has given rise to horny scales or other protective coverings. Even in these cases, however, scattered hairs are present, and in the aquatic forms referred to, bristles occur about the lips of the young. So the hairy covering is as characteristic for the Mammalia as the horny scales are for the reptiles. Although, like scales and feathers, hairs are epidermal structures, nourished by a papilla of the corium, yet there is a fundamental difference between them in their development. Both scales and feathers begin by a thickening of the epidermis which projects beyond the level of the skin, and, in the case of the feather, is only afterwards retracted into the follicle, but the hair begins by a thickening of the epidermis which grows inwards into the cutis, and only afterwards comes to project beyond the level of the skin. (Fig. 100.)

In the Cat two kinds of hairs are present, those which constitute the fur, and those which form the whiskers (*vibrissæ*); the latter from the richness of their nerve-supply are especially tactile in function. In many of the Carnivora and other orders a soft under-fur is overlaid by stronger bristle-like hairs,

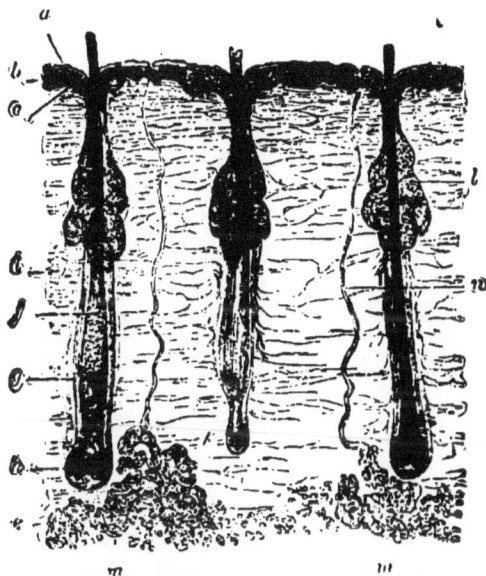

Fig. 100—Section through skin of horse—enlarged. *a*, epidermis ; *b*, its Malphigian layer, *c*, papillary layer of corium, *d* ; *e*, Subcutaneous tissue ; *f*, hair in its follicle, *g*, with papilla *h* ; *i*, old hair being replaced by *k* ; *l*, sebaceous glands ; *m*, sweat glands ; *n*, sweat duct.

which form the external coat. The hairs are lubricated by the secretion of the **sebaceous** glands, which open into the necks of the hair-follicles. The skin of the Mammal is therefore richer in glands than is that of the Reptiles or Birds. In addition to these, however, there are also **sweat** glands, which select from the blood certain materials which have to be excreted from it, and so the skin of the Mammal comes to be an important excretory organ. Aquatic Mammals alone are destitute of these glands. Of the two kinds of glands referred to, it is the sebaceous kind which the milk-glands resemble most as to structure.

5. In all Mammals (with the exception of some aquatic forms) it is possible to distinguish all the five regions of the vertebral column—cervical, dorsal, lumbar, sacral, caudal. In some aquatic forms (Fig. 101) the hinder extremities have so nearly

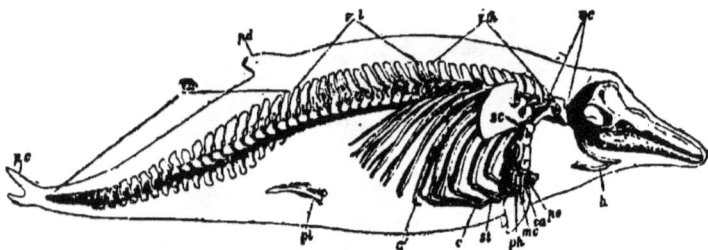

Fig. 101. Skeleton of Porpoise. (*Phocœna*). vc, cervical vertebræ anchylosed ; vth, dorsal ; vl, lumbar ; vx, caudal regions; pd, dorsa.., pc, caudal fins ; pl, pelvic rudiments ; c' false, o true ribs ; st, sternum ; sc, scapula ; h, humerus ; r, radius ; ca, carpus ; po, pollex ; mc, metacarpus ; ph, phalanges.

disappeared that it becomes impossible to distinguish a sacral region, but the ribs are always present and connected to the sternum in such a way as to mark off the cervical, dorsal and lumbar regions. As we advance through the Vertebrate series we see a tendency towards the reduction in number, and towards a constancy in the reduced number of certain parts. The teeth, for example, will furnish us with a good illustration of this rule, but it is also well exemplified in the cervical region of the vertebral column, for not only is it much shorter than usual in the Reptiles, but (with the exception of a few genera which also in other respects present primitive characteristics) it always contains seven vertebræ, and that in spite of the extremely long and short necks which we meet with in different members of the class. The other regions vary in different orders as to the contained number of vertebræ, within narrower or wider limits, the caudal most of all ; the Cat, *e.g.*, has thirteen dorsal, seven lumbar, three sacral, and twenty caudal vertebræ. (Fig. 102). Instead of the sternal ends of the ribs being ossified as they are in birds, they are always cartilaginous, and

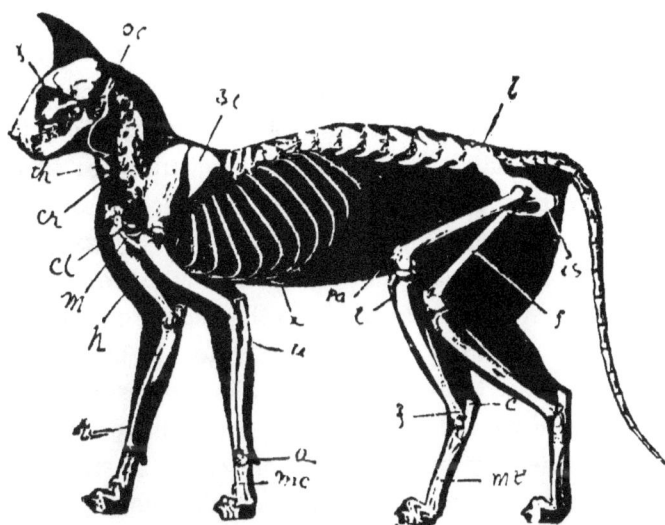

Fig. 102.—Skeleton of Cat. (After Strauss-Durkheim.)

oc, occipital ridge; sc, scapula; il, ilium; is, ischium; f, femur; pa, patella; t, tibia; fi, fibula; c, calcaneum; mt, metatarsus; mc, metacarpus; a, pisiform; r, radius; n, ulna; h, humerus; x, xiphoid cartilage; m, manubrium sterni; cl, clavicle; cr, cricoid; th, thyroid.

the sternum presents more traces of its original mode of formation than it does even in Reptiles. Thus, in the Cat, it is composed of eight pieces, the first of which, the **manubrium**, carries the clavicles in the mammals in which these bones are complete, while the last, the **xiphoid** cartilage, does not become ossified like the other pieces.

6. The skull of the cat is, of course, modified in connection with the strong teeth and muscular development of the jaws, but apart from such superficial characters, a knowledge of its structure will enable the student to understand that of the other mammals, and to compare it with that of the lower forms. It is first to be observed that the bones of the cranium and face, with the exception of the mandible, are immovably united together, as in the turtle or bird, but the sutures remain quite distinct, and the bones can be readily separated from each other in a macerated skull. The first thing that strikes one **in comparing**

the skull of a cat with that of a lower Vertebrate is the relative proportion of the cranium to the face. Here it is obvious that the facial bones form only a small portion of the skull, the increased size of which, on the other hand, is chiefly due to the large surface of the frontal, parietal and squamosal bones which form the greater part of the brain case. A second feature is the widely-arched form of that bar (**zygomatic** or **jugal**), which extends from the articulation of the lower jaw to the superior maxilla, and is, therefore, comparable to the quadrato-jugal bar of the reptiles. It is widely arched, partly to accommodate underneath it the great **temporal** muscle, which arises in the temporal fossa bounded behind by the strong lambdoid crest, and partly to give a wider origin to another muscle of mastication which arises from its convex outer surface, the **masseter.** A third feature is the tendency to union of certain groups of bones separated by sutures in the reptiles ; thus the mammalian occipital bone is formed of the four occipital elements and rests on the atlas vertebra by two condyles which are borne on the exoccipitals ; it likewise absorbs in older animals an unpaired **interparietal,** which is wedged between the supraoccipital and the parietals. The mammalian **temporal** bone is similarly complex, because it embraces not only the three periotic bones with the tympanic and squamosal, but probably also the quadrate and quadrato-jugal. Within the bullate tympanic is to be found the chain of small bones answering to the columella auris of the reptiles. A third complex bone is the **sphenoid,** the elements of which, however, are more easily separated from each other. They are the basi-, pre-, ali- and orbito-sphenoids with the pterygoids. Finally the **ethmoid,** which looks into the skull with its perforated plates, into the nasal cavity with the septum and the labyrinthine turbinals, and into the orbits with its orbital plates, is formed of mesethmoid and parethmoids.

Attention should also be directed to the following points :—

the union of the frontal and jugals, the strength of the jugal processes of the maxillaries, and of the alveolar parts of these bones, the hard palate with the incisive foramina between the nasal and mouth cavities, and the concealment of the vomers by the palatines, the foramina for the escape of the various cranial nerves, that in the lachrymal bone for the passage into the nasal cavity of the tear-duct, and, finally, the intracranial aspect of the various bones, with the ossified partition (**tentorium cerebelli**) which separates the cerebrum from the cerebellum, and the hollow on the upper surface of the basisphenoid (**sella turcica**) for the accomodation of the hypophysis.

Certain muscular processes on the lower jaw, especially the coronoid process to which the temporal muscle is attached, indicate the powerful character of the parts concerned. The articular surface of the mandible is convex, and elongated transversely; it fits into a corresponding concavity (**glenoid**) bounded behind by a ridge (**postglenoidal**) on the root of the zygomatic process of the temporal bone. It is a question whether this part of the temporal bone is comparable to the quadrate of the lower forms or not. The fact that the Meckelian cartilage is continuous in development with one of the chain of bones within the drum of the ear (the **malleus**) caused anatomists to believe that, in comparison with the Sauropsida, the Mammalian mandible has shifted its articulation to the squamosal, but this continuity is not irreconcilable with the interpretation of the glenoid region of the temporal as the quadrate.

The complex visceral skeleton of the fish is only represented in the Mammal by the hyoid bone and its two pairs of cornua. Both of these are attached to the extremities of a curved basal piece, the anterior representing the hyoid, the posterior the first branchial arch of the lower forms. The former is attached to the temporal region of the skull, and often unites with it forming a bony process (**stylo-hyoid**) of the temporal bone. **In the**

rest of its course it may be ligamentous, or formed into cerato-
and epi-hyal bony pieces. The posterior cornua are formed by
one bony piece on each side, which on account of their supporting the larynx receive the name of **thyro-hyal**.

7. Considerable difference will be observed in the structure
of the shoulder-girdle, and the mode of its attachment to the
trunk. In the Sauropsida we saw that an intimate connection
of the skeleton of the anterior limb to the trunk was secured by
the union of the clavicles and coracoids to the sternum. But
in the Mammals both these bones may be absent or very much
reduced in size, the shoulder-girdle being represented only by the
scapula, which is then connected to the trunk by muscles alone.
In the cat *e.g.*, the coracoid is in the form of a mere process
projecting from the scapula in the neighborhood of the glenoid
cavity, and the clavicle, which in man is connected to the spine of
the scapula and the manubrium of the sternum, is rudimentary.
As regards the rest of the appendicular skeleton little need be
said, except to call attention to the comparatively primitive
arrangement of the parts (with the exception of the position of
the radius), to the incipient or complete absence of the first toe,
the greater development of one element in the carpus and tarsus for the attachment of muscles, and the fact that the metacarpals and metatarsals (**metapodials**) are not in contact with
the ground in locomotion.

In regard to the first point it will be observed that the radius,
although related to the inner border of the carpus and hand, is
connected with the outer border of the humerus ; this is attributable to a twist in the lower end of that bone, which changes
the position of the radius at the elbow joint, and renders it
necessary that it should cross the ulna so as to reach the inner
border of the hand.

In many members of the cat's own order, so-called plantigrade
Carnivores, the feet have a larger surface of contact with the

ground than in the cats, which are on this account called digitigrade. The terminal joints of the toes in the cat-tribe are attached in such a way, that, when not in use, the claws which they bear do not touch the ground; they are retracted into a sheath and are thus always kept sharp. In all the forms, the skin underlying the portions of the feet which touch the ground is converted into pads or balls, the thickened epidermis of which affords protection to the underlying cutis.

8. A comparsion of the relative weight of the brain and body in the cat with that in any of the lower classes of Vertebrates will convince us that a great advance in intelligence is to be expected, and, indeed, the relative size of the brain-case and facial region in the skull already discloses its superiority in this respect. At first sight it may be difficult to compare the cat's brain with that of any of the lower forms studied, for certain of the regions which are visible in the reptile or bird are here concealed by the overgrowth of other regions. This is especially true of the thalamencephalon and the optic lobes, which are entirely hidden by the backward extension of the cerebral hemispheres. The cerebellum also has gained in size, especially its lateral lobes, which are joined by a series of fibres crossing transversely under the fore part of the hind brain, and constituting the pons Varolii. In front of the pons, the fibres which ascend through the medulla oblongata towards the cerebrum diverge in two masses, the crura, to the hemispheres, and in front of the crura is the only part of the thalamic region which reaches the surface of the brain, viz. the hypophysis and infundibulum. On the whole, the most important changes are those which have taken place in the cerebrum, for not only has it increased in size by growing forward, backward and towards the sides, but the grey matter of its surface, instead of being smooth, is folded inwards in such a way as to leave a series of fissures on the surface of the cerebrum with intervening convolutions. We

have already seen in the cerebellum of the bird a similar plan for accommodating a large s irface of grey matter in a comparatively small space, and the cat's cerebellum also exhibits the characteristic "arbor vitæ" arrangement. The cerebral hemispheres are not independent of each other, for, apart from certain transverse bundles of fibres present in the lower classes, there is also formed between those parts of the hemispheres which project back over the true roof of the brain (Fig. 103) an important transverse commissure, the **corpus callosum**, which serves to effect communications between the two sides. In some higher

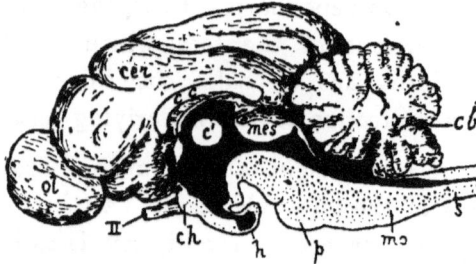

Fig. 103.—Median longitudinal section of brain of cat (Modified after Wilder.)

s. Spinal cord with contained central canal. *mo*, medulla oblongata ; *p*, pons Varolii ; *cb*, cerebellum, forming the floor and roof of the 4th ventricle. *mes*, optic lobes or corpora quadrigemina, the roof of the mesocœle or " iter," the floor is formed by the crura in front of the pons. In front of the mid-brain is the thalamio region, the cavity of which (3rd ventricle) is traversed by a commissure, *c*, uniting the side walls or thalami ; the roof is partly formed by the epiphysis (above *mes*), and the floor by the hypophysis *h* ; but the chiasma, ch. of the optic nerves, II, also forms part of the floor, while the anterior thin wall (terminal lamina) is above the chiasma. The cerebral hemispheres, *cer*, are right and left of the lamina, and connected by *cc*, the corpus callosum ; *ol*, the olfactory lobe

mammals the cerebral hemispheres project so far forward as to conceal the olfactory lobes from above, but in the cat these are well marked off from the rest of the brain.

9. As far as the sense-organs are concerned, we shall find that the nose and ear are those which exhibit conspicuous advantages in structure over the lower classes. The olfactory cavity in the fowl is not entirely smooth, but in the cat the olfactory mucous membrane covers an extremely complicated surface, furnished chiefly by the ethmoid bone, but also supported by independent turbinal bones. The olfactory lobes rest upon the cranial surface of the mesethmoid, separated by a median crest in that bone, and sending the olfactory nerve-fibres down-

ward into the mucous membrane of the complicated labyrinth, through the perforated or sieve-like plates of the ethmoid. The right and left nasal cavities are separated by the perpendicular plate of the ethmoid, and by the cartilaginous partition (*Septum*), which extends as far as the nostrils. Certain small cartilages not present in the lower forms are developed in the external nose for its support, and for the attachment of the muscles which move the nostrils. An organ of Jacobson, such as is well developed in the snakes, is present in the cat, but it has only a cartilaginous capsule, and it opens into the mouth on each side through the incisive foramina. In the far-back position of the posterior nostrils, the cat and other mammals resemble such reptiles as the turtles and crocodiles, but the pterygoid bones do not take part in bounding the nostrils.

10. Reference has been made above to the great development of the tympanic cavity, but there are also other modifications of the auditory apparatus, which make the ear of the mammal more efficient (Fig. 104). In the first place the outer ear, or **pinna**, is formed with its supporting cartilages, which serve for the attachment of the muscles which move it. The chain of bones, also, which effects communication between the tympanic membrane and the labyrinth, is more complex ; it contains three elements, the malleus, incus and stapes, of which the first is connected in the young with Meckel's cartilage while the last, a stirrup-shaped bone, occupies the fenestra ovalis. Zoologists at one time thought that from the connection of the malleus with Meckel's cartilage, the former bone must be comparable to the quadrate of lower forms, but the view is gaining ground that the whole chain is comparable to the columella auris of the lower forms.

With regard to the labyrinth, an important advance in structure, which we meet with in most mammals as compared with the Sauropsida, is that the cochlea, instead of being a straight

11

tube, is coiled up like the coils of a snail's shell (from which indeed the name cochlea is derived), and the branch of the nerve which goes to it has to occupy the axis or columella of the coil, so as to reach the whole length of the tube.

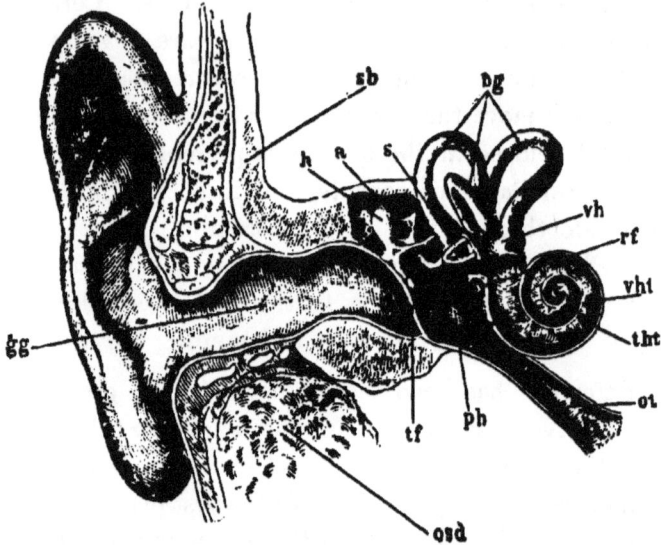

Fig. 104.—Partly diagrammatic representation of auditory apparatus in man.
(The external and middle ear are in their proper position, but the labyrinth is rotated inwards through 90°; both tympanic cavity and labyrinth are twice the natural size, the external ear is the natural size). gg, auditory passage; tf, tympanic membrane, ph, tympanic cavity; ot, Eustachian tube; h, malleus; a, incus; s, stapes; vh, vestibule of labyrinth; bg, semicircular canals; vht, scala vestibuli of cochlea; tht, scala tympani, leading to fenestra rotunda, rf; sb, temporal bone; osd, parotid gland.

11. The mammals present just as great diversity in the nature of their food and in their method of securing it as do any of the lower classes. Important reactions on the structure of the creature, especially on its intestinal system, are to be expected, and the teeth as well as other parts are thereby affected. Much importance is attributed by systematists to these organs, because they are readily accessible to inspection, are the only parts of the intestinal system preserved with the skeleton in fossils, and are extremely constant, not only in number but also in form, for any particular species. Like most mammals, the cat

has two sets of teeth, the milk set and the permanent set: it is **diphyodont**. Some mammals are **monophyodont**, they have only one set of teeth, which are destitute of the fangs or roots of ordinary teeth, and are simply added to in length by the pulp, as their free surface becomes worn down. Other forms, which are diphyodont with regard to the greater number of their teeth, still retain some with this unlimited power of growth. Although in some reptiles certain teeth are distinguished from their neighbours either by their form or by standing isolated in the series, yet no such specialisation of the teeth in the different parts of the gape occurs, such as we find in the mammals. Here the premaxillary bones lodge teeth of a distinct form, the **incisors**, which in the cat are six in number, and are separated by a gap or **diastema** from the large and sharp **canines**, the foremost teeth of the maxillaries. Corresponding but alternating with these are similar teeth in the lower jaw, but the gaps are in this case behind not in front of the canines. Behind the canines are the **premolars**, *i.e.* the teeth which replace the back teeth of the milk dentition, and behind these the true **molars**, which only appear in the permanent set. Both kinds of grinders are reduced in number in the cat in accordance with the principle of specialisation referred to in § 5.

There are three premolars and one molar in the upper jaw but only two of these are functional, the first premolar and the molar being evidently rudimentary. On the other hand, there are two premolars and one molar in the lower, all three of which are functional teeth. The same alternation which is to be seen further forward in the gape is also to be seen here, so that the cat's dental formula of one side might be expressed thus :—

$$\frac{\text{i i i c pm pm } \textbf{PM} \text{ m}}{\text{i i i c } * \text{ pm pm } \textbf{M}}$$

The teeth marked in full-faced type are the characteristic

teeth of the carnivorous dentition, the canines and **sectorials**, specially developed in accordance with the carnivorous habits, while the asterisk indicates the gap caused by the absence of the first premolar.   It is interesting to compare this formula with the less specialised dentition of a dog:—

$$\frac{\text{i i i c pm pm pm PM m m}}{\text{i i i c pm pm pm pm M m m}}$$

Here it is obvious that the last premolar above, and the first molar below are, as in the cat, the sectorial teeth, and although the molars proper are more numerous, they show the same tendency to a rudimentary character at either end of the series.   As far as numbers go, the bear's formula is the same as the above, but there is no specialisation of the sectorial teeth, and both the bear and the dog present a formula, which is very near to what is generally considered the **typical formula** of the Mammalia, only differing therefrom in the absence of a third upper molar.   Rewriting, then, the cat's dental formula after comparing it with that of the dog, we should have the following better expression of its dentition.

$$\frac{\text{i}^1\ \text{i}^2\ \text{i}^3\ \text{c} \ *\ \text{pm}^2\ \text{pm}^3\ \textbf{PM}^4\ \text{m}^1\ *\ *}{\text{i}^1\ \text{i}^2\ \text{i}^3\ \text{c} \ *\ *\ \text{pm}^3\ \text{pm}^4\ \textbf{M}^1\ *\ *}$$

When it is not considered necessary to indicate the relative position and character of the teeth, the dental formula is generally simplified ; thus, (for the cat):— i$\frac{3}{3}$ c$\frac{1}{1}$ pm$\frac{3}{2}$ m$\frac{1}{1}$.

In respect to their form, the molars in the Carnivores hardly deserve the name of grinders, for the functional molars have sharp cutting edges not adapted for grinding. The bears and their allies, however, which are more omnivorous in their habits, show how a cuspidate molar may be worn down to a grinding surface.

12. In most mammals the lips and tongue attain considerable independence of movement ; this is associated with the **development** of a complicated musculature which is hardly

represented in the lower forms. The hinder part of the roof of the mouth likewise is contractile, and becomes the soft palate, a flap of which, known as the **uvula**, hangs down in the middle line, so as to cut off a posterior chamber or **pharynx** from the true mouth-cavity. The aperture between the two is further narrowed by folds at either side, the pillars of the **fauces**, between which are the **tonsils** belonging to the lymphatic system. (I, 63). The pharynx communicates above with the nasal cavities, below with the larynx and œsophagus, the aperture into the former tube being always protected by the **epiglottis**, a movable fold behind the base of the tongue. The primitive mouth-cavity which we studied in the lower forms is thus not only separated into a respiratory chamber above (the nasal cavities), but into an alimentary chamber below, which further presents from before backwards, a buccal cavity between the lips and the gums, the mouth-cavity proper and the pharynx. The structural advance which we thus see in the mammals is partly determined by the use of the lips for prehension, partly by the longer retention of the food in the mouth, for admixture with the secretion of the salivary glands, for mastication and for submission to the organ of taste.

13. In all mammals the salivary glands are arranged in three masses, parotid, submaxillary, and sublingual, and their secretion partly serves to facilitate the swallowing of the food, and partly to aid in the digestion of the starchy constituents thereof. The organ of taste is composed of certain bud-like structures which recall the sense-organs of fishes (I, 9), and which are chiefly situated in the walls of trenches surrounding the large **circumvallate** papillæ at the back of the tongue. Fungiform and filiform papillæ are also present in the tongue of all mammals, the latter clothed with horny epidermic sheaths in the carnivores, and giving the front of the tongue its rasp-like surface.

14. In the lower forms studied, the cœlom is an undivided

cavity, in front of which is the heart with its separate serous
sac, the pericardium ; the lungs, when present, project into the
cœlom, and are covered with the same serous coat as the intestines, but, in the mammals, the cœlom is sub-divided by a
muscular partition (the **diaphragm**), into a thoracic cavity containing the lungs and the heart, and an abdominal cavity containing the other viscera.   That part of the cavity which
belongs to the lungs is now called the pleural cavity, the heart
with its pericardium remaining independent between the two
pleuræ.   The diaphragm acts chiefly as a muscle of respiration,
serving by its construction to enlarge the thoracic cavity, and
thus to facilitate the changes of the air in the lungs.   Certain
structures like the œsophagus, the aorta, and various nerves,
must pierce the diaphragm so as to reach the abdominal cavity ;
they do so after traversing the **mediastinum**, or space between
the pericardium, the pleuræ and the vertebral column.   The
intervention of the diaphragm thus forms a sharper boundary
between the œsophagus and stomach than we have met with
in the lower forms.   In no mammal has the stomach the
simplicity met with in the Ichthyopsida or Reptiles.   If it be
considered as a dilatation of the intestinal tube, it is always a
one-sided dilatation, so that there is a short curvature (the un-
dilated side of the tube), and a greater curvature (the dilated
side).   Its form in the Cat is quite simple, but in herbivorous
animals there is generally a much more complex stomach, and
in these also the intestine is relatively much longer.   The chief
difference in the mammal's intestine as compared with lower
forms, is the proportionately much greater length of the large
intestine.

    15. As regards the respiratory system, apart from the differ-
ences referred to above, the most important is the more min-
utely " cellular " character of the lungs, the bronchi dividing
up dichotomously into smaller tubes which eventually end in
the alveoli or air-cells.   There are thus no membranous por-

tions of the lung-walls or sac-like projections thereof, such as occur in the Sauropsida. The upper part of the windpipe is of great interest, however, because it is first in the Mammalia, in which we meet with the characteristic arrangement of the hard parts of the larynx, which attain their perfection in the vocal organ of man. In addition to the cricoid and arytænoid cartilages, such as are met with in the lower forms, a third, the thyroid, is developed which forms the "Adam's apple" of the human larynx and serves with the arytænoids for the attachment of the vocal cords. A membrane stretches from the outer edge of the thyroid cartilage to the thyrohyals (§ 6), and the larynx is thus brought into connection with the hyoid bone.

16. We saw that the great respiratory activity of the birds requires a complete separation of the arterial and venous blood in the heart : this is effected in the same way in the mammals, the right auricle and ventricle, respectively, receiving the venous blood from the body and pumping it through the lungs, while the left auricle and ventricle, respectively, receive the ærated blood from the lungs and pump it through the system (Fig. 105). Some difference is to be observed in the great vessels which come off from the heart, notably as to the great aortic arch, which is the fourth arterial arch of the left, not of the right side, as in birds. The blood which

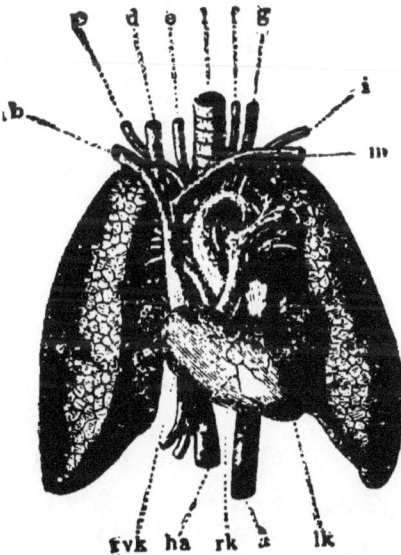

Fig. 105.—Heart and Lungs of Man
a, descending aorta ; lk, left, rk, right ventricle ; rvk, right auricle ; ha, inferior vena cava ; l, trachea ; e and f, the two carotids ; beside these, d and g, the two internal jugular veins ; c and i, subclavian arteries ; b and m, subclavian veins.

returns from the intestines is subjected to the action of the hepatic cells while passing through the portal circulation, but the venous blood from the posterior extremities is returned directly to the heart through the posterior vena cava, the kidney (which first attains in the mammals its reniform shape) receiving its blood supply entirely through the renal arteries.

17. Having now examined into the structure of one of the typical mammals, or Eutheria, let us see in what respects the Prototheria and Metatheria differ therefrom.

The Prototheria embrace only a single order—**Monotremata**—represented by two well-known forms, the Duck-mole (*Ornithorhynchus paradoxus*), and the Porcupine Ant-eater (*Echidna hystrix*) of the Australian region (Figs. 106 and 107). Out-wardly, and in their habits, these creatures differ very much from each other, the Duck-mole being an aquatic animal, with soft un-der-fur covered by

Fig. 106.—The Duck-billed Platypus. ‡
(*Ornithorhynchus paradoxus*).

stiff over-lying bristly hairs, which prevent the wetting of the fur, with webbed feet which adapt it for swimming, and with a horny, toothless bill like a duck's, evidently adapted to secure food in the same way ; while, on the other hand, the Porcu-pine Ant-eater is pro-vided with stout bur-rowing feet, by the aid of which it opens the

Fig. 107.—Porcupine Ant-eater. (*Echidna hystrix*). ⅓ ants' nests, on the con-tents of which it feeds, is protected by stout spines instead of the bristly coat of the Duck-moles, has a sharp snout

like most of the ant-eaters, and secures its food with its tongue. There are no teeth, at least rudimentary teeth which are present in the embryo never break through the jaw, so we are obliged to consider the absence of teeth not as a primitive character, but as an adaptation to habits of life.

Inwardly, however, the Monotremes agree with each other, and differ from the other mammals in many important respects, such as the presence of an episternum (III. 18), with which the clavicles articulate, and of complete coracoids, which unite the scapula and sternum. There are also other features which, like their oviparous habits and the structural characters associated therewith, remind us more of the Sauropsida than the Mammalia ; such are the simplicity of the brain and of the ear, and the temperature of the blood, which is much more like that of the surrounding medium than in the ordinary mammals.

18. The Metatheria similarly correspond to a single order, the **Marsupialia.** This order receives its name from the pouch referred to above (§ 2), which is supported by two epipubic bones, present in all members of the order, whether the pouch is well-developed or not. The Monotremes also have these epipubic bones, so it is probable that they are not formed in connection with the pouch, but rather represent such structures as the fore part of the pelvis in the Menobranch. Although the Marsupials are specially developed in the Australian region, they are not exclusively confined to it, for in America there are several species of opossum (*Didelphys*), and in the Oriental region also, there are a few representatives of the group. Australia is, nevertheless, the home of the Marsupials of the present day, although it is of interest to note, that the earliest fossil remains of mammals, obtained in Europe and other parts of the world, indicate that they ought to be associated with the Metatheria instead of the Eutheria, and, therefore, that this sub-class was at one time much more widely distributed than it now is.

A great variety of forms occurs among the Australian Marsupials. There are some like the Tasmanian devil (*Dasyurus*), or like the native dog, (*Thylacinus*) which are distinctly carnivorous in their habits ; other fruit-eating forms which are arboreal, and are adapted for their mode of life by the possession of a long prehensile tail, or even, as in the case of the flying phalangers (*Phalangista*), of a patagium like our flying squirrels ; again, there are herbivorous forms like the kangaroos (Fig. 108), in which the fore legs are extremely short, the hind limbs chiefly used in locomotion and the toes reduced in number, and, finally, there are forms (*Phascolomys*) with gnawing teeth like the beaver's, which are associated with similar methods of securing food. Such features as the above are evidently adapted to the habits of the creatures, but there are

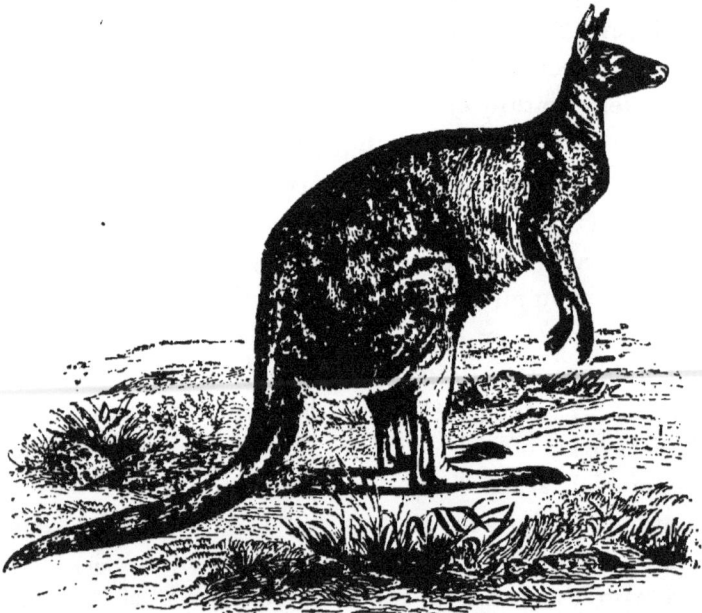

Fig. 108. Giant Kangaroo. (*Halmaturus giganteus*). ⅛

certain underlying structural peculiarities common to the whole group apart from those refered to in § 2. The dentition e. g., is not referable to the same type as that of the higher Mammalia, the teeth being much more numerous ; the tendency to union of different tracts of skullbones (§6), is not so well-marked, and the angle of the lower jaw bone is turned in, in a characteristic way, which has assisted in the identification of the fossil remains of this nature.

19. When we finally study the group of the higher mammals or Eutheria, we find a wonderful diversity of form in the different orders, depending on their habits and methods of locomotion. Certain aberrant orders may first be referred to, which occupy a somewhat isolated position in the sub-class. Of these the **Bruta** or **Edentata** is a very heterogeneous order, embracing the ant-eaters of the Old and New Worlds, and the sloths of South America. The former differ very much in the clothing of the skin, for in the Indian genus *Manis* (Fig. 109), it is formed of large overlapping horny scales, while in the South American *Myrmecophaga*, coarse hair replaces these. Both genera have a long snout and a long protrusible tongue by means of which (and the secretion of the large salivary glands) they secure their

Fig. 10). Scaly Ant-Eater or Pangolin.   (Manis longicaudata). $\frac{1}{8}$

food. The Brazilian armadillos (*Dasypus*) and some other South American allied forms have the skin of the back and sides converted into a more or less complete shield of bony plates, while the African *Orycteropus* is clothed with coarse hair. Unlike the Carnivores, the teeth, if they are present at all, are all alike, often very numerous, and there is only one set.

20. Contrasting with these forms which have all strong burrowing feet are the sloths, in which the claws are curved in such a way as to be only useful for an arboreal life. The teeth are less numerous than in the insect-eating armadillos, and their surfaces are flat and not tuberculate ; the toes in accordance

with their different function are reduced to three (*Brady-pus*) (Fig. 110, F) or two (*Cholœpus*). Very complete remains of extinct forms intermediate between these two subdivisions of the order have been found in South America; these include gigantic forms like *Megatherium*, almost as large as an elephant, which probably fed on the foliage of trees, uprooted by their powerful limbs.

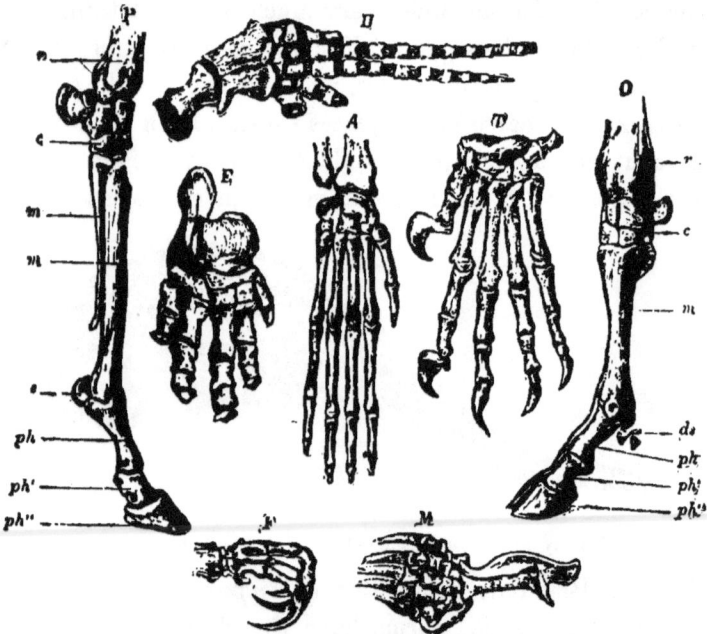

Fig. 110. Manus of various Mammalia.
P, Horse. D, Dolphin. E, Elephant. A, Orang. T, Tiger. O, Ox. F, Sloth.
M, Mole. r, radius. c, carpus. m, metacarpus. s, sesamoid bone.
ph, phalanges. ds, dew-claws.

21. A second aberrant order is that of the **Cetacea**, which owe their peculiarities to their aquatic life. This is not the only order of Mammalia in which aquatic habits are present, for certain carnivorous animals like the seals, sea-lions and walruses, are exclusively or almost exclusively confined to water, yet in the Cetacea, this adaptation is carried so far as to isolate them from the other orders of the sub-class to which they belong.

The general spindle-shape of the body (Fig. 111), the hairless skin, the thick layer of blubber, the rudimentary character of the olfactory organ, and the consequent restriction of the nostrils to the respiratory function, the situation of the nostrils on the top of the head, the conversion of the anterior extremities into flippers (Fig. 110, D), the almost complete absence of the skeleton of the posterior extremities, the peculiar horizontal caudal fin, and the dorsal fin occasionally present, are all features which are asso-

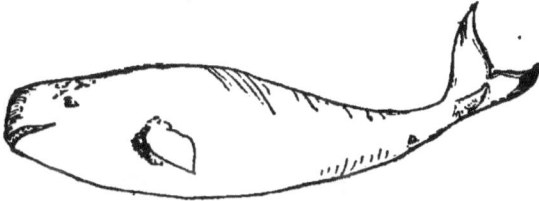

Fig. 111. Outline of White Whale of St. Lawrence. (*Beluga.*) ₁ₜₒ.

ciated with the conditions of their existence. The whales have only one set of teeth, but these disappear in some of the members of the order, being replaced on the upper jaw by the horny strainers (whalebone), which prevent the escape of the minute creatures on which the whalebone-whales live. Two groups of Cetacea are therefore distinguished—the toothed whales and the whalebone-whales. To the former belong the porpoises, (*Phocæna*); dolphins, (*Delphinus*); white whales, (*Beluga*, Fig. 111); as well as the grampus (*Orca*) and the singular Narwhal (*Monodon*); they chiefly live on fish and cuttle-fish, for seizing which they are provided with numerous sharp conical teeth, but the Grampus has only a few very powerful and sharp teeth, in accordance with its habit of attacking the larger forms of its own order, and the adult Narwhal is quite toothless, except for the single long spiral tusk of the male.

Only the lower jaw is provided with teeth in the Spermaceti whales (*Catodon*), which are chiefly remarkable for the enormous head swollen up by the accumulation of spermaceti between the skull and the skin.

A considerable proportion of the length of the body in the whalebone-whales likewise belongs to the head. The members of this group attain the largest size of any whales, some of those with a dorsal fin (*Physalus*) measuring as much as one hundred feet in length, while the right whales, which are the chief objects of the whale fisheries, never measure more than sixty feet. (Fig. 112.)

Fig. 112.—Outline of Greenland Whale. (*Balæna mysticetus.*) ₁₆₀

22. A third aberrant order, that of the **Sirenia,** have a certain superficial resemblance to the Cetacea, which is associated with their aquatic mode of life, but they are not so completely adapted thereto, being herbivorous forms, and thus necessarily frequenting the shallow waters of the shore-zone for sea-weeds, or ascending the estuaries of the great tropical rivers and browsing upon the vegetation which fringes their banks. The manatees (*Manatus*) of Western Tropical Africa, and of Eastern South America, are, indeed, not completely helpless on land, the fingers in the flippers are marked by short nails, and in all the forms, the presence of a neck, the position of the nostrils at the end of the muzzle, and the less rudimentary character of the pelvis indicate less departure from the typical mammalian form than is to be seen in the Cetacea. One of the forms, the northern sea-cow, (*Rhytina*), exterminated little more than a century ago, was previously abundant on the shores of Siberia and Kamschatka. It was toothless, the mouth being provided with four horny-toothed pads, which served in place of the grinders of the living forms. These are more numerous in the manatee, than in the dugong of the Indian

Ocean (*Halicore*, Fig. 113), but the latter has tusk-like incisors in the upper jaw, which are only transitorily present in the manatee. The caudal fin of the latter is by no means so effective a propelling organ as is that of the dugong, which creature is, on the other hand, quite helpless on land. While the skin in the sea-cow was extremely thick and hairless, that of the manatee is covered with stiff bristles, which are both fewer in number and shorter in the dugong. Some fossil

Fig. 113.—Dugong (*Halicore.*) ₁/₁₀

members of the order are known, in which both the teeth and the skeleton of the hind limbs are more completely represented than in the living Sirenia, but these, instead of uniting the group to the Cetacea, rather prove an alliance with the hoofed animals, to the study of which we now proceed.

23. The remaining orders of Mammalia arrange themselves naturally in two series, the Hoofed Animals (**Ungulata**) on the one hand, those provided with claws and nails on the other (**Unguiculata**). Although, at first sight, this distinction appears to be of little importance, the hoof being a horny covering for the whole of the distal joint of a toe, while the claw or nail is merely developed on one surface (the anterior), yet it is the mark of a difference of function which is associated with some of the most characteristic peculiarities of the Ungulata. In by far the greater number of the living hoofed animals the extremities are devoted entirely to the function of locomotion, and in most cases the number of toes is reduced in accordance with a

principle referred to above. They thus contrast with the Unguiculata, which retain for the most part the typical number of toes. If we, however, extend our survey from the living to the fossil Ungulates, we shall find that the reduction of the number of toes is comparatively recent in the history of their series, the oldest forms being five-toed like most Unguiculata.

24. The most primitive order of the hoofed animals in this respect (Fig. 110 E) is that of the **Proboscidea** or Elephants, although in other respects it is one of the most specialised and highly organised of the series. The only living genus is *Elephas*, represented by two species *E. indicus* and *africanus*, the latter distinguished from the former by its enormous ears, and by the lozenge-shaped ridges on the molar teeth. In both the living forms, the thick skin has only a few isolated bristles, but the fossil mammoth (*E. primigenius*), which was abundant in

Fig. 114—Skeleton of Mastodon.

Siberia and Alaska, and of which frozen carcases have been found, was covered with wool intermingled with coarse hair. Most of the peculiarities of the skull are attributable to the singular dentition of the Elephant, which consists of two huge incisor tusks in the premaxillaries and, in addition, six grinders

on each side in each jaw, which, however, instead of being present simultaneously, succeed each other, so that only one, or at
most two of the six are in function at the same time.   The length
of the tusk-sockets causes the great height of the skull, which   .
is especially large in the Indian elephant, not so much from the
size of the brain, as from the great thickness of the middle or
spongy layer of the cranial bones.   In the fossil genus *Mastodon*,
(Fig. 114) of which many remains are found throughout Ontario,
there were tusks in the lower jaw as well ; the grinders suc-
ceeded each other as in the elephant, but they had from three to
six transverse rows of tubercles, from which the genus derives its
name.   One can arrive at the structure of the elephant's tooth
from that of the mastodon, by imagining the transverse rows
added to in number, their surfaces worn down and the intervals
between them filled up with cement.

The neck is so short that the proboscis or trunk, an elongated
external nose with a finger-like process at its tip over the nostrils,
becomes necessary for securing food ; it requires, therefore, to be
very muscular and sensitive. The apparent disproportion between
the length of the fore and hind limbs is due to the fact that the
latter are inclosed within the skin of the loins nearly to the
knee-joint, but the bones of the fore and hind legs are approximately of the same length.   A curious difference of gait
is observable between the elephant and higher Ungulates on
account of the position of the elbow and knee joints in the former;
they occupy the middle of the limbs, the metapodials being
quite short and contributing to the formation of the soles of
the feet, whereas in the latter, the metapodials become the
long cannon bones, the wrist and ankle joints are raised to
the middle of the limbs, being known as the " knee " and
" hock " respectively, and the true elbow and knee joints are
close up to the trunk.   The higher Ungulates therefore are not
plantigrade like the elephants, but walk on the tips of the
distal joints of the digits.   (Fig. 110, P and O).

12

From what has been said above, it will be understood that the geographical distribution of the elephants was formerly by no means so limited as it is at the present day ; for, in addition to the mastodons already referred to, true elephants were abundant in Ontario during the pleistocene period, and ranged southwards through the States to Mexico.

25. Associated with the earliest fossil elephants, there have been found in Miocene strata in Europe remains of a proboscidean of gigantic size (*Dinotherium*), the skull of which measures some five feet in length, and differs from the elephant's, in that some five molars replace the single grinder, and that the lower jaws contain the tusks. The conformation of the skull suggests a trunk like the elephant's, and the bones of the feet, which have also been found, prove it to be referable to this order. No examples of it have been found in America, but in the earlier Eocene strata are found remains of mammals of the size of elephants, and with bones so similar, that some nearer alliance is suggested than that they are mere predecessors in time. Their skull and teeth, however, were specialised in a very different direction from that of the Dinotherium, for the former was provided with three pairs of bony cores for the attachment of horns (whence the name of the principle genus and the order **Dinocerata**), and the latter were arranged in the following peculiar way :—

$$i\,\tfrac{0}{8}\ c\,\tfrac{1}{1}\ pm\tfrac{3}{3\,\text{or}\,4}\ m\tfrac{3}{3}$$

there being no upper and very small lower incisors, while the upper canines were tusk-like, the lower ones small, and the small molars with two transverse ridges. Casts of the cranial cavity of these early Mammalia show that the brain was very small in size, and low in its type of structure.

More primitive than Dinoceras and its allies with regard to the teeth are *Phenacodus* and *Coryphodon* from the lower Eocene, which have the typical formula § 11, and, in the former case, truly tuberculate teeth. All the toes are present, the middle one in Coryphodon, however, being distinctly the longer, as in the tapirs. No such primitive hoofed animals persist till the present day, all of them (with the exception of the elephants) having undergone a reduction of the number of toes.

26. Before we proceed to the typical hoofed animals there is one aberrant genus (*Hyrax*) to be mentioned, which is placed in an order by itself (**Hyracoidea**). This order includes several species of timid little crea-

tures of the size of rabbits, which extend from the Cape of Good Hope to Syria (the coneys of Scripture), living in crevices in rocky and mountainous districts. The toes are four on the fore, and three on the hind limbs, being all provided with flat hoofs, except the inner hind toe, which is clawed. Unlike those of the higher Ungulates, the hoofs do not support the weight of the body, which rests on the soft soles attached to the under surface of the other joints of the toes, and which, applied sucker-like to the rocks, enable them to perform marvels of climbing. Like the rabbits they have a rodent dentition, $i.e.$, the incisors—$i\frac{1}{1}$—wear down to a chisel-shaped edge and have growing roots, the upper are curved, the lower horizontal, there are no canines and the molars ($pm\frac{4}{4}$ $m\frac{3}{3}$) are provided with transverse ridges. In other respects the Hyrax is more nearly related to the true Ungulata, and especially to the order of the odd-toed Ungulates, which we now proceed to study.

27. Reference was made to the fact that in one of the earliest fossil Ungulates the middle (third) toe is longer than the others, and therefore contributes more to the support of the body. A similar preponderance of this toe is to be observed in all odd-toed(imparidigitate) Ungulates, which constitute the order called on this account **Perissodactyla.** The first and most primitive family of these is the *Tapiridæ*, represented by one Indian and two South American species of Tapir (*Tapirus*). These are swamp-loving creatures with short smooth hair, a short tail, an almost trunk-like proboscis, feet four-toed in front and three behind like the Hyrax, but with a more primitive dentition, $i\frac{3}{3}$ $c\frac{1}{1}$ $m\frac{7}{7}$. Although so limited in number at the present day, numerous fossil tapirs are known, as well as forms (*Palæotherium, &c.*), connecting them with the Coryphodons, and with the Rhinoceroses which constitute the second family of the order. This family (*Rhinocerotidæ*) contains only a single genus with four species living in India, Java, Sumatra and Africa respectively, the two former being one-horned, the two latter two-horned species. In all, the skin is provided very sparingly with hair, is very thick and often divided off into shields; the horns are not supported by a bony core. In the living forms there are only three toes (viz. nos. 2, 3, 4) present, but in certain

fossil forms the fifth toe was also present in the fore foot, as in the Tapir. The earlier fossil Rhinoceroses had a more complete set of teeth than the living species, which have no canines, and have a tendency to lose the incisors, while the molars are present to the full number $\frac{7}{7}$. Like the Elephants, the Rhinoceroses had once a much wider geographical distribution, for remains of the woolly rhinoceros are found with those of the mammoth in Siberia, and numerous American representatives of the family have also been found. Among these are Miocene forms which had two horns side by side, and attained elephantine dimensions *(Brontotherium)*.

28. The third family of Perissodactyla is represented at the present day by the single genus *Equus*, to which the horse and various species of asses and zebras belong. It differs from the foregoing, in that the body is supported entirely on the third toe, the distal joint of which (coffin-bone) is covered with the hoof, while the second joint (coronary) and proximal joint (fetter-bone), but especially the metapodials (cannon-bones) are much elongated, bringing the wrist and ankle-joint up to the middle of the leg (Fig. 110 P). The second and fourth toes have disappeared almost entirely, but they are represented by the rudimentary metapodials (splint-bones), the proximal ends of which only are complete. Occasionally it occurs that a horse is born with these rudimentary digits in a more perfect condition, the splint-bones not only being complete, but carrying short digits, the ends of which may be clad with miniature hoofs. Such a three-toed horse is evidently a reversion to a more primitive Perissodactyle type, and such reversions are known as instances of " atavism."

This family presents a more typical dentition than does the foregoing. The incisors are $\frac{3}{3}$, the canines small, especially in the mare, and there is a long diastema between the front teeth and the grinders which number $\frac{6}{6}$, the first milk molar not reappearing in the permanent dentition. A peculiarity of the incisors is that the surface enamel is folded in like the inverted finger of a glove, the result being a ring of enamel which constitutes the mark of the incisors, until, in the aged horse, the tooth has been worn down below the fold. In all the members of the family, the hair of the mane and tail is long, and there are present callosities in the skin near the knees and hocks, but in the asses and zebras, the hair is only long at the tip of the tail, and the callosities are only present on the fore-legs. The zebras are South African forms distinguished by black stripes on a cream-coloured ground; the asses occur in North Africa, Western and Central Asia, the North African species being probably the source of the domestic ass. Although horses are found in a feral condition (*i.e.*, apparently wild, but really only secondarily so) in Asia and South America, it is uncertain whether the original stock still exists in a wild condition; some recent investigations, however, in the high table-lands of Thibet point to the conclusion that such is the case. The South American horses were imported by Europeans, but it is not to be supposed that the New World was until the time of its discovery uninhabited by horses. Fossil remains of true horses show that they were abundant in America long before their importation from Europe, and from the various Tertiary strata numerous representatives of the family have been discovered, so that the American fossil Equidæ are much more numerous than the European forms. Those from the lower Tertiaries had both a more complete dentition and also a greater number of toes, recalling in this respect the genus *Coryphodon* (§ 25) ; but as we study the forms which occur in the higher Tertiary strata we find a gradual loss first of the fifth, then of the second and fourth toes until the third alone is left with the rudiments referred to above. The earlier Equidæ were small, about the size of a fox, but they gained in size during the Tertiary period until they attained the stature of the horses of the present day.

29. The great bulk of the Ungulates belong to the order **Artiodactyla,** in which the third and fourth toes equally support the weight of the body.   Here we have also primitive forms, and specialised forms adapted for rapid locomotion ; the primitive forms being, as in the last group, swamp-loving or aquatic creatures, with comparatively hairless skin, and with teeth much more nearly approaching the tuberculate type than do those of the more specialised forms.   As far as the dentition is concerned, the swine are the most primitive of the living species, but the Hippopotamus is supported by all four toes, and therefore, in this respect, it is the more primitive form.

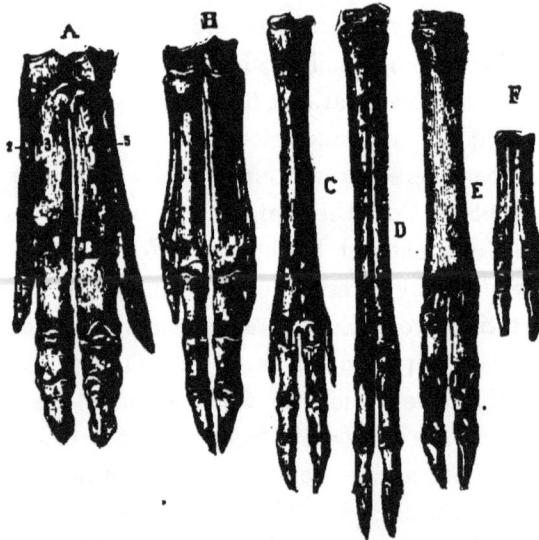

Fig. 115.—Reduction of the lateral toes, and coalescence of metapodials into a cannon-bone in Artiodactyla. (After Gaudry.)
   A, pig ; B, *Hyæmoschus* ; C, roe ; D, antelope (*Calotragus*); E, sheep; F, embryo calf.

These constitute two families, the *Hippopotamidæ* and the *Suidæ,* which differ from the remaining Artiodactyla in not chewing the cud. The first family has only a single genus, which, at the present day, is represented by one or two species in the rivers of Africa. Their dentition is singular, for although the milk teeth are i$\frac{3}{3}$, c$\frac{1}{1}$, m$\frac{4}{4}$, the adults have only i$\frac{3}{2}$, c$\frac{1}{1}$, m$\frac{6}{6}$, there being no permanent successors for one of the milk incisors and one of the milk molars. The permanent front teeth are tusk-like, the incisors being straight, while the canines are curved, and meet each other in such a way, that the posterior faces of the much larger lower ones are ground flat against the anterior faces of the upper ones.

The second family has many more representatives, and a much less limited geographical distribution, for there are several Old World genera, as well as the Peccaries, which are peculiar to the New World. The skin in all is more or less closely beset with bristles, their bodies are more elongated, and thus better adapted for rapid locomotion, and they are supported solely by the third and fourth toes ; the second and fifth, although they are complete and furnished with hoofs, not reaching the ground. (Fig. 115—A.)

The Peccary even offers a further reduction in the hind foot, for the fifth toe there is undeveloped. This genus (*Dicotyles*) is in many respects the most specialised of the family, for apart from the structure of the hind foot, and a reduction in the number of incisors and molars, the stomach resembles in its complexity that of the Ruminants. On the other hand, the genus *Sus* is the most primitive, for its dentition is i$\frac{3}{3}$, c$\frac{1}{1}$, pm$\frac{4}{4}$, m$\frac{3}{3}$, whereas in the other genera, there is either a reduction in the number of the incisors or molars, or both. The canines are generally tusk-like, the lower ones being the chief weapons in the family, but the upper ones may also attain a formidable size, as in the pig deer—Babyrussa—of the Moluccas (*Porcus*), where they are curved upwards and backwards; the incisors are small and sometimes absent in the adult, as in the pigmy hog (*Porcula*) of India, and the wart-hog (*Phacochœrus*) of Africa, but the molars are always of a tuberculate pattern.

30. The foregoing families constitute the non-ruminant section of the Artiodactyla; all the other numerous genera are **ruminant** forms, the stomach being complex, so as to admit of their characteristic way of feeding. This and the reduction of the second and fifth toes are both to be regarded as subservient to the more rapid locomotion in this group, for these herbivora, which are the chief objects of pursuit by the larger carnivores,

are able to secure and to bolt, without previous mastication, in a very short time, a large amount of food, which they after-wards masticate when they have got to some secure retreat. The mechanism concerned may be studied in the sheep's stomach, where the cardiac end has two compartments, the larger Rumen (*paunch*), and the smaller Reticulum (*honeycomb*), while the pyloric end has similarly two compartments, the Psalterium (*manyplies*), and the Abomasum (*rennet stomach*). The œsophagus is attached between the rumen and the reticulum, and the grass which is hastily swallowed passes first into these compartments; it is then moved from one to the other, and finally thrown back into the mouth and subjected to a thorough mastication and insalivation, after which the semi-fluid product is again swallowed and strained off into the abomasum and in-testine, through the psalterium, which is connected indirectly with the œsophagus by a half-groove on the wall of the reticulum, capable of being converted into a complete channel.

A peculiar dentition accompanies the ruminant stomach; in the typical forms it is $i\frac{0}{8}$, $c\frac{0}{1}$, $m\frac{8}{8}$, there being only a pad in the upper jaw, against which the lower incisors and incisor-like canines bite. A wide gap separates the front teeth from the molars, which have flat crowns with semilunar folds of enamel on the surface. Such teeth are therefore said to belong to **selenodont** forms, in contradistinction to the tuberculate teeth of **bunodont** forms, but it is obvious (as in the Hippopotamus, *e.g.*,) that a tuberculate tooth when worn down may present a peculiar pattern of enamel, and there-fore bunodont forms are regarded as more primitive (Fig. 116).

Fig. 116.—Molars from the upper jaw of Bunodont and Seleno-dont fossil Artiodactyls, *Palæo-chœrus* and *Xiphodon*.

The feet are also different in their structure in the ruminant forms, for not only are the second and fifth toes raised off the ground, becoming **dew-claws**, but they often disappear, and the weight of the body rests entirely on the tips of the third and fourth toes, being transmitted to them through a **cannon bone**, which is formed by the more or less complete coalescence of the third and fourth metapodials (Figs. 115 and 110 O). There are generally horns in this group, often confined, however, to the male sex; when there are no horns, tusk-like canines may serve as compensatory weapons of de- fence. The sheep, oxen, antelope and deer are the typical ruminants, but there are some aberrant families which we may shortly consider first.

Of these the *Camelidæ* present some peculiar features ; for example, the psalterium is absent; the upper jaw in front is not destitute of teeth, for it retains one of the upper incisors and the canine on each side, which are absent in the ruminants : the canines, the upper especially, are tusk-like and the molars are only ⅔ Instead of walking on the tips of the digits, they are digitigrade forms, all three joints resting on the ground ; the hoofs are thus of no use in locomotion, a single or double pad of skin being present on the under surface of the third and fourth digits. There are no dew-claws, but the third and fourth metapodials have not so completely coalesced as in the oxen. The geographical distribution of the family has the same peculiarity which we have noted of certain other groups, there being both Old and New World representatives, quite isolated from each other at the present day. In the Old World the genus *Camelus* only occurs in a domesticated condition, as the Dromedary and the Bactrian Camel, the former in Arabia, Africa, and India, the latter (the two-humped camel), in the Mongolian table-lands. Both are used as beasts of burden, and it is likely that the humps are the result of their employment as such. There is only a single cushion underneath the digits, which, therefore, present a suitable surface for the sandy soil of the desert. In the New World, on the other hand, there are both wild and domesticated forms belonging to the genus *Auchenia* ; these are smaller sized forms without a hump, they tread on a double pad, have fewer molars and larger ears than the camel. The wild species are the larger Huanaco, and the smaller Vicuña ; the domesticated forms, the Llama (the beast of burden of the mountain regions of Peru), and the Alpaca, which is kept in herds for its flesh and its wool. The explanation for the existence of the two branches of this family is to be sought in the Miocene Strata of America, where numerous remains of Camelidæ are found, with complete front teeth, and separate metapodials. The descendants of these must have made their way into the Old World by a bridge which existed between North America and Asia, and which afterwards subsided so as to form Behring Straits.

Like the *Camelidæ*, the pigmy deer of Java and West Africa *(Tragulidæ)* have tusk-like upper canines in the males and only three compartments of the stomach ; their metapodials are less completely coalesced than in the camels. In size and in the possession of tusk-like canines, the musk-deer of Central Asia *(Moschus)* resemble the pigmy deer, but the male has a peculiar musk-gland on the skin of the ventral surface.

A fourth group is that of the Giraffes *(Camelopardalis)* an African form much nearer the typical Ruminants than any of the foregoing. They differ from them both in the general form, and in the horns, which are skin-bones covered with soft skin, the most primitive of the horns of the ruminants.

32. It is according to the nature of these structures, that we subdivide the typical Ruminants into the hollow-horned *(Cavicornia)* and the antlered forms *(Cervidæ)*. In the former, projections of the frontal bones form the so-called cores, which are covered with the variously-shaped but usually unbranched horn, in the latter, the frontal bones bear the antlers, with an intervening ring of bone, the "rose," where the antlers are broken off each year. The antlers are only covered by skin (the velvet) while they are undergoing formation ; that process complete, the velvet is rubbed off, and the polished bone exposed. We noted that, in several of the preceding families, only the males have tusk-like upper canines ; in these forms, also, it is of common occurrence that the males only should have the weapons of offence and defence in the shape of horns or antlers. Among the *Cervidæ*, the Reindeer or Caribou *(Rangifer)* is singular in that both sexes have antlers, which like those of the Moose *(Alces)* are broad and palmated in form. Those of the Stag *(C. Canadensis)* as well as of the Virginia deer *(Cariacus virginianus)* carry rounded branches.

Among the *Cavicornia*, the Prong-horn of the Western prairies (*Antilocapra americana*) is the only form with a tendency to branching of the core. It comes nearest the *Cervidæ* in this respect, as well as in the fact that it casts its horns at intervals. It is usually classified with the Antelopes, a sub-family of the *Cavicornia*, whose headquarters are Africa, but which are less numerous in Europe and Asia. The only other American Antelope is the Rocky Mountain goat (*Haplocerus americanus*). A well-known European genus is the Alpine chamois (*Rupicapra*), while the gazelle (*Antilope dorcas*) is a common North African species. Some of the African antelopes approach the next subfamily, the oxen (*Bovina*), in their proportions. This group embraces the domestic ox (which includes various races probably derived from several wild species), the humped zebu (*Bos indicus*) and two or three other Indian species. Other genera are the Old World buffalo (*Bubalus*), the European and American bison, the yak of Thibet (*Poëphagus*) and the musk-ox of the Arctic regions (*Ovibos moschatus*). A third subfamily (*Ovina*) embraces the sheep, which are found wild on high mountain ranges, e.g., the bighorn (*O. montana*) of the Rocky Mountains and the Argali of Central Asia, also the ibexes (*Capra ibex*) and goats (*C. hircus*). Our domestic sheep are probably derived from one of the Asiatic forms.

33. We must now turn to the Unguiculate orders, looking in the first place somewhat more closely into the classification of the Carnivora, one form of which has been already studied.

Of the six families recognized in this order, the *Ursidæ* is least specialised; it embraces plantigrade forms, with a dentition in which the sectorial teeth are not prominent, and which are frequently omnivorous. Examples of these are the kinkajou of Brazil with its prehensile tail (*Cercoleptes*), the racoon (*Procyon lotor*) and the true bears (*Ursus*). Nearest to these are the badgers (*Taxidea*) and skunks (*Mephitis*), which form a plantigrade section of the weasel family (*Mustelidæ*), to which there also belong the chief fur-bearing animals of North America, the martens (*Mustela*), minks and ermines (*Putorius*), wolverines (*Gulo*), otters (*Lutra*), and sea-otters (*Enhydra*). The third family (*Viverridæ*), embracing the civet cats and the ichneumons (*Herpestes*), is chiefly an Old World family, as is that of the *Hyænidæ*, but the dogs (*Canidæ*) and cats (*Felidæ*) are abundantly represented on both continents, the former embracing the dogs, foxes, wolves and jackals; the latter, the lions, tigers, leopards, lynxes and cats of the Old World, and the pumas, ounces and lynxes of the New.

34. A very interesting branch of the Carnivora is that of the Pinnipedia often placed in an independent order, and much modified in accordance with their aquatic mode of life. There is a marked tendency in this group, which contains the Seals,

Walrus, etc., towards the reduction in number of the incisor teeth ; the canines are rarely tusk-like, and the extremities are converted into flippers, the hinder ones being turned backwards parallel with the short tail (Fig. 116). The seals proper are those

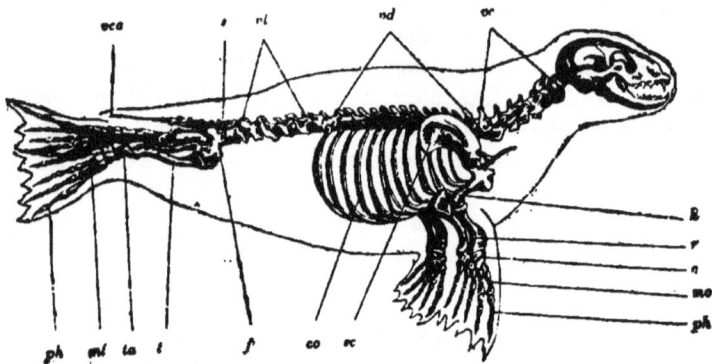

Fig. 116.—Skeleton of Seal.

vc. cervical, vd. dorsal, vl. lumbar, s. sacral, vca. caudal regions of vertebral column; h. humerus; r. radius; c. carpus; me. metacarpus; ph. phalanges; sc. scapula; co, ribs.

which are least adapted for locomotion on land ; the walruses and sea-lions, on the other hand, can raise themselves off the ground by the aid of their limbs, and the flippers of the sea-lions even have divisions, corresponding to the toes. As a rule there is no external ear, the head is rounded, and the cylindrical body diminishes in girth towards the tail.

Of the three families distinguished, that containing the eared-seals (*Otariæ*), is nearest such Carnivora as the sea-otter, the structure of the flippers and their habits suggesting a less perfect adaptation to an aquatic life than we meet with in the seals proper. In most of the species the coat is formed of stiff, bristly hairs alone, but in several, such as the Alaska fur-seals (*Callorhinus ursina*), the bristles are scarcer, and there is a thick, soft under-wool. These skins, when dried and the bristles removed, yield the valuable seal-skin furs, but the skin of the seal is also sought after for other purposes, and the blubber of all the species yields a valuable oil. Some of the Otariæ, such as the California sea-lion (*Eumetopias Stelleri*), reach the large size of fifteen feet. The other species are smaller.

The *Phocidæ* chiefly differ from the above in the absence of external ears and the shape of the flippers, those of the hinder extremities being especially remarkable on account of their notched outline, due to the shortness of the middle and the length of the inner and outer toes. The harp seal of the St. Lawrence is one of the commonest species (*Phoca grœnlandica*, Fig. 195), but there are some singular forms also in this family, such as the hooded seal of the North Atlantic (*Cystophora cristata*), and the gigantic sea-elephant of the Antarctic Ocean, (*C. proboscidea*), which attains a length of twenty feet, and is distinguished by the prolongation of the nose into a proboscis. Some species of the family live in land-locked seas, such as the Caspian and the lakes of Newfoundland.

The Walrus (*Trichechus rosmarus*) is placed in a family by itself, characterized by the enormous upper canine tusks, by the shape of the trunk, which does not diminish in girth towards the loins, by approaching the Phocidæ, in respect to the absence of external ears, and the Otariæ, in the ability of the creatures to raise the body on the limbs, and thus leave the sea, and even climb steep rocks so as to reach a safe place above high water, where they may bask in the sun.

It is observed that the young sea-lions take somewhat unwillingly to water, and swim at first awkwardly ; this is one indication, among many others, that the Pinnipedia are a group of mammals which have gradually acquired aquatic habits, and with them their modified form. The same is thought to be true of the Sirenia, for fossil forms have been found allying them to the terrestrial Ungulata. Great uncertainty is felt by zoologists, however, as to the alliances of the third group of aquatic mammalia—the Cetacea ; many, nevertheless, think that these carnivorous aquatic forms may have been originally more seal-like in form, and that the horizontal caudal fin had within it the hard parts of the hinder extremities, of which hardly a trace is now to be detected ; no fossils have yet been found to confirm this idea, although various anatomical considerations render it probable.

35. In turning to the other orders of Unguiculate mammals, we shall find some forms that exhibit more primitive features than the Carnivores, others that are specialised for an aerial, arboreal or subterranean life, as much as the Pinnipedia are for a life in water.

36. As far as regards the dentition, the **Insectivora** are certainly the most primitive ; they are also all plantigrade forms, and have well-developed clavicles. The teeth are not more numerous than in the Carnivores, but the canine tooth does not assume the function which it has in most mammals, sometimes an incisor, sometimes a premolar, project-ing further from the jaw than it does itself. Seven molars are by no means always present, but their surfaces are always tuberculate, in accordance with the insect-food.

Between groups of Rodentia and Insectivora, which corres-pond in their habits, a certain superficial resemblance is to be detected, a convergence of character which we attribute to the influence of their surroundings, but the Rodents, as we shall learn, including about one-third of all the species of mammals, offer a greater wealth of form than we find in the Insectivora. To these two orders belong all the smaller species of Mammalia, and, indeed, parallel groups of both orders are to be be found, some adapted for a life on the surface of the ground, some for burrowing underneath it, some for a semi-aquatic, and some for a more or less completely arboreal life.

Only two out of the six families are represented in our region, the Shrews (*Soricidæ*), and the Moles (*Talpidæ*). The former are mouse-like Insectivora with an elongated muzzle, and a short velvety coat. In both the common forms, the eyes are small, and in one (*Sorex platyrhinus*) the ears and tail are long, while, in the other (*Blarina brevicauda*), both are short. Both of these are terrestial species, but an aquatic genus (*Myogale*), in which the toes are webbed and which has a very penetrating musky odour, is found in South-east Russia and the Pyrenees. In all, the hinder feet are larger than the fore, but, in the Talpidæ, the fore feet are converted into broad shovel-like structures, with short toes and stout claws for digging the burrows in which they live (Fig.110M); the limbs are very short, the body elongated and cylindrical in outline, the head very small, without either evident eyes or ears. The shape of the body is ob-viously adapted to the underground life, and the snout is provided with extremely delicate tactile sensibility. This is best seen in the Star-nosed moles (*Condylura cristata*), where the fleshy disc at the end is divided up into projecting rays; in the other genera, *Scalops* and *Scapanus*, the Common and Hairy-tailed Moles, the snout is simply pointed.

The other families are confined to the Old World; their typical genera are *Erinaceus*, the European hedge-hog, a terrestrial form in which the hairs are converted into stout spines; it protects itself by rolling itself into a ball, and thus causing the spines to diverge from each other. In another family, to which the *Centetes* of Madagascar belongs, the spines are replaced by bristles, which, with the pointed snout and strong lower canines, give the creature some resemblance to a miniature pig. The Malayan genus *Tupaja* and its allies are arboreal insectivores with the soft fur and habits of squirrels, and, in the same region, a singular genus, *Galeopithecus*, is to be met with, provided with a patagium like our flying squirrels. It is the representative of an independent family, as is also the African *Macroscelides*, marked by the length of the hind limbs, like the Jerboa, and occurring, like it, in rocky and desert regions.

37. The chief contrast between the Insectivora and Rodentia is in the nature of the food, and the difference in structure and habit brought about thereby. Nowhere is the difference in structure more evident than in the dentition, where the incisors are reduced in number, grow from persistent pulps, and acquire a chisel-shaped edge, from having the enamel only on the anterior surface. Generally the incisors are only $\frac{2}{2}$, but in the hares and rabbits (*Lepus*), there is a small tooth behind each of the curved upper incisors; the canines are always absent, and the molars never tuberculate, but provided with transverse folds of enamel. As in the Herbivora, the stomach is constricted into cardiac and pyloric chambers, and each of these may have recesses; further, the alimentary canal is long in proportion to the body.

Nearly half of our N. American Mammalia are Rodents belonging to seven families, the *Leporidæ*, *Hystrichidæ*, *Muridæ*, *Dipodidæ*, *Geomyidæ*, *Castoridæ*, *Sciuridæ*. The first of these includes the hares and rabbits, and is sufficiently characterized by its dentition. The type of the second is the Old World Porcupine (*Hystrix*), represented in N. America by the common porcupine (*Erethizon*). Both of these forms have spines, which are more efficient weapons of defence than those of the hedgehog,

on account of the ease with which they can be detached from the skin.
To the *Muridæ* belong the rats and mice (*Mus*), as well as the field-mice
(*Arvicola*) and the aquatic musk-rat (*Fiber*). An interesting northern
genus is the lemming (*Myodes*), which often migrates in vast numbers
from one part to another in Northern Europe and Asia. The *Dipodidæ*
include the Egyptian Jerboa (*Dipus*), marked by the length of the hind
legs ; the same peculiarity is present but less developed in the American
jumping mouse (*Zapus*). In the Western prairies, two genera of pouched
gophers are met with, which constitute the family *Geomyidæ*. They have
cheek-pouches which open on the cheeks outside the mouth, and are lined
with hair. As the gophers are burrowing forms, the fore feet are large
and armed with strong curved claws. The most truly aquatic of the
Rodents is undoubtedly the beaver (*Castor fiber*), the largest of our
species. As in the musk-rat, the hind toes are elongated and webbed, and the flat, scaly tail is very characteristic. The last family
—the *Sciuridæ*—includes several genera, chiefly arboreal forms with soft
fur, and more numerous molars than the preceding, except the hares.
Among those which are distinctly terrestrial in their habits, may be
mentioned the gophers and prairie-dogs, while the chipmunk and the
woodchuck are intermediate, in this respect, between these and the true
arboreal squirrels and flying squirrels. The completely terrestrial forms
have cheek-pouches which open into the buccal cavity ; these are best
developed in the gopher (*Spermophilus*) and the chipmunk (*Tamias*),
while they are rudimentary in the prairie-dog (*Cynomys*), and absent in
the woodchuck (*Arctomys*) and the squirrel (*Sciurus* and *Sciuropterus*).
The last-mentioned genus includes the nocturnal flying squirrels, which
have a patagium stretching from the fore to the hind limbs and permitting a slanting leap, such as we have already observed to be possible in
several mammalian orders.

As representatives of tropical families, may be mentioned the Chin-
chilla of Chili and Peru, valued for its grayish fur, the Guinea-pig
(*Cavia cobaya*, now only known in the domesticated condition), the Capy-
bara (*Hydrochærus*, the largest Rodent), the Paca (*Cœlogenys*) and Agouti
(*Dasyprocta*) of Northern South America, which four genera have almost
hoof-like nails, and, like most of the Rodents, are gregarious in their
habits.

38. On the approach of cold weather many animals of different classes pass through a resting phase, which, in the warm-
blooded animals, is usually spoken of as hybernation. This

phenomenon occurs in several orders of Mammalia, but nowhere is it more easily studied than in the Rodents. During this period the various functional activities are arrested as much as possible; there is no food taken and little tissue consumed, so that, as respiration is also very inactive, the body temperature is assimilated to that of the surrounding medium, and thus the warm-blooded animals become temporarily cold-blooded.

39. After the Rodents, the greatest number of mammalian species belong to the bats—**Chiroptera**—our next order. Here we have a very different modification for aerial life than we have hitherto met with; instead of a patagium like that of the flying squirrels, Galeopithecus and the Phalangers, certain of the fingers and the fore-arm are much elongated and serve to spread a very delicate hairless web, which extends backward to the thighs and frequently also surrounds the tail (Fig. 117). The thumb

Fig. 117.—Outline and skeleton of Phyllostoma (after D'Alton).
oa, humerus; va, radius and ulna; w, carpus; I, pollex; II—V, second to fifth fingers; m and f$^{1\ 2\ 3}$, elongated metacarpals and phalanges; zf, femur.

alone does not take any part in the support of the web, this being chiefly effected by the third, fourth, and fifth fingers. Except in the fruit bats, it is the only digit which bears a claw; in these, however, the index also is clawed, although, like the other fingers, it is entirely enveloped in the web. The web is very sensitive

13

and as the bats are nocturnal creatures, the nerve-terminations in it constitute one of the chief channels through which sensations reach the brain. In further accordance with their nocturnal habits, the bats have small eyes, and large ears; they hybernate in cold climates, where they are often to be found in large numbers, hanging in some cave or building, by the claws of the hind feet, wrapped up in the patagium. The bats are characteristically insectivorous forms, very few are fruit-eating, but the dentition of the two groups indicates a sharp contrast between them. As the patagium does not merely serve to break the force of a fall or to permit of an oblique leap, but is a true organ of flight, the pectoral muscles require to be specially developed to permit of such use of the anterior extremity, and consequently the sternum is provided with a crest, and the clavicles are stronger than they are elsewhere among mammals. It is not surprising that the bats should be the most widely distributed of Mammalia.

The fruit-bats are abundant in eastern tropical countries, and attain the largest size of any members of the order. *Pteropus*, the fox-bat, so called on account of the pointed snout, may be mentioned as a type of this section, the fruit-eating habits of which are indicated by the blunt tubercles of the molars. In the insectivorous bats, on the other hand, the tubercles of the molars are sharp or coalesce into a W-shaped cutting edge. The snout is short, and the external ears are of large size. Two groups are recognized—those in which the external nose is provided with a membranous expansion round the nostrils, and those in which there is no such membranous expansion. To the former belong the vampire-bats of S. America (*Desmodus*), which attack and suck the blood of horses and mules. With that exception, they are mostly insect-eating forms, but the genus *Vampyrus* of Guiana lives chiefly on fruit. The ordinary bats (*Vespertilionidæ*), with the nose destitute of the membrane, are represented by two genera in our region, as examples of which may be mentioned the little brown bat (*Vespertilio subulatus*) and the red bat (*Atalapha noveboracensis*).

40. As the Australian continent is peopled by a remarkably primitive mammalian fauna, so also the Island of Madagascar possesses characteristic mammals which are found nowhere else,

and which occupy, in some respects, a comparatively low place among the unguiculate Eutheria. With few exceptions the mammals of Madagascar belong to the order Prosimii, and the members of this order are also, with few exceptions, confined to Madagascar and the neighbouring parts of the Continent of Africa. As well the fact, however, that there are certain outlying members of the order in the Malay Archipelago and India, and that fossil remains have been found in various parts of the world indicate that their geographical distribution was not always so restricted. They are completely arboreal forms, the inner digits of both fore and hind feet being opposable, and thus forming thumbs. On this account, they were long associated with the monkeys under the ordinal name Quadrumana, but it is more convenient to consider them apart from the monkeys, although they are undoubtedly allied in some respects to them. Their dentition is peculiar, the incisors being $\frac{2}{2}$, or reduced in number, the canines absent in a rodent-like genus (*Chiromys*), and the molars tuberculate like those of the Insectivora, but they do not confine themselves to insect food, living also upon smaller Vertebrates, fruit, etc. The second digit of the hind limb is always clawed, while the other digits bear nails, such as those of the monkeys. Most of them are nocturnal creatures which have a soft warm coat, and often a bushy tail.

Two families are formed for the reception of the aberrant genera *Tarsius* and *Chiromys*. The former is found living socially in the woods of Borneo, and has received its name from the great length of the tarsus. It is also singular on account of the enormous size of the eyes. The *Chiromys* is the Aye-Aye of Madagascar, a form which picks out larvæ from the trees on which it lives, by means of an extraordinarily thin finger (the third), which it inserts into their burrows. Its rodent-like incisors suggest that its food is not confined to larvæ.

The third family *Lemuridæ* includes all the other genera, some of which are very bizarre creatures. The Loris (*Stenops*, &c.) somewhat resemble the Spectres (Tarsius) in their distribution and habits. The Galagos are carnivorous creatures which vary from the size of a rabbit to that of a mouse, while the Lemurs proper are of large size, attaining in the case of the tailless Indri *(Lichanotus)* nearly three feet in length. The other Lemurs are smaller in size and provided with a long tail which is coiled about them for warmth, while they rest during the day. In *Propithecus* the snout is short as in the Indri, the result being a monkey-like face, while in the ring-tailed lemur *(L. catta)* and its allies, the snout is prolonged.

41. However monkey-like certain of the lemurs are, they form a decidedly more primitive group than that of the monkeys proper. This is especially noticeable in the structure of the brain, the cerebellum being left uncovered by the cerebrum in the former group, while, in the latter, the hinder lobe of the cerebrum is so developed as to overlap the cerebellum entirely. In this respect, as well as many other anatomical features, the monkeys agree in structure with Man, and, accordingly, they are generally placed together in the order **Primates**, in spite of the exceptional place which Man otherwise occupies in nature. In all Primates the incisors are ⅔, the inner digits (thumb and great toe) are opposable (except in man, where this is only true of the thumb), and all the fingers are nailed, not clawed. The orbits, which have complete bony walls, are directed forwards, and the face, in comparison with the Lemuroids, is hairless.

Fig.118—The Uakari.--*Brachyurus calvus.*
As an example of the Platyrrhini.
(From Brehm.)

Apart from Man, three families of Primates are recognized, two of which are New World groups. Most of the S. American monkeys belong to the **Platyrrhini**, so called on account of the width of the septum of the nostrils, which causes these apertures to look outwards. The tail is usually prehensile, assisting in their arboreal life in the dense forests which they inhabit; they differ from both the other families in having an extra molar tooth, and a rounded skull. Examples of this group are the howling and spider monkeys (*Mycetes* and *Ateles*), the bonnet-monkeys (*Cebus*), and certain smaller squirrel-like forms, with soft, abundant fur, and nocturnal habits, which depend upon their feet alone for climbing (*Nyctipithecus, Chrysothrix*). The remaining South and Central American forms are called **Arctopitheci**; they have one true molar less than the Platyrrhines, tuberculate grinders, and fingers with claws instead of nails. Only one genus is recognized (*Hapale*), including several species of marmosets, the smallest of the Primates.

The Old World monkeys (**Catarhini**) have a thin nasal septum, the nostrils directed downwards and forwards, the same dental formula as Man, (i⅖, c⅟, pm⅖, m⅜) and the tail, if present, never prehensile. There are two groups of them, those that approach Man (*Anthropomorpha*) in the absence of the tail and of cheek-pouches, as well as in the less prominence of the face, and the greater length of the anterior limbs, and those that approach the Carnivora (*Cynomorpha*) in the strength of the facial region, and the development of tusk-like canines, while they differ from the other group in having shorter anterior limbs, and, very often, cheek-pouches. To the Anthropomorpha belong the chimpanzee and the gorilla of Western Africa (*Troglodytes*), the orang-utan (*Simia satyrus*) of Borneo and Sumatra, and the gibbons (*Hylobates*) of the same islands and the continent of India. To the Cynomorpha belong forms generally of smaller size, such as the sacred monkey of the Hindoos (*Semnopithecus*), the African *Colobus*, the squirrel-like *Cercopitheci* of Africa, the macaques. (*Macacus*—chiefly Asiatic, with the exception of the tailless macaque, *M. ecaudatus*, of North Africa, which is preserved in Gibraltar), and the baboons, which are the most dog-like in face of the *Cynomorpha*, and include, among the African species, some very large forms.

CHAPTER VII.

THE ARTHROPODA.

1. It was stated in Chapter I., that the term Invertebrata includes several distinct sub-kingdoms, resembling each other in that they do not possess the arrangement of the nervous system and skeleton typical for the Vertebrates. It must not be understood that all are equally unlike Vertebrates, some worms for example, seem to foreshadow in their structure the vertebrate organization, but the most highly organized Invertebrates—the **Arthropoda**—diverge very widely from the type of structure which we have studied heretofore, although their organs are built up of histological elements, similar, in many respects, to those of the Vertebrates. The nature and extent of the divergence referred to may be studied in the crayfish, an Arthropod which forms a suitable introduction to the sub-kingdom to which it belongs, both on account of its size and on account of its position in the group.

2. The Arthropoda are bilaterally segmented animals like the Vertebrates, but, unlike them, their segmentation is visible on the surface of the body, especially on account of the fact (from which the group derives its name) that each segment may carry one pair of jointed appendages. Throughout the sub-kingdom the rule holds good which obtains also in the Vertebrates, that, in the more primitive families, the segments are not only more numerous but less constant in number, and show less tendency to be grouped into dissimilar regions. The nervous system partakes in the segmentation of the body, but is situated on the ventral aspect, while the centre of the blood-vascular system is dorsal, so that there is a complete reversal of the neural and hæmal aspects, as compared with the Vertebrates (Fig. 119) ; nevertheless, it is the neural aspect which is first developed in both sub-kingdoms. No endoskeleton affords attachment to the muscles of the body or protects the delicate organs, but this is functionally replaced by an exoskeleton of chitin, a hard sub-stance of peculiar chemical composition, secreted by the skin, and sometimes rendered harder by the admixture of calcareous salts.

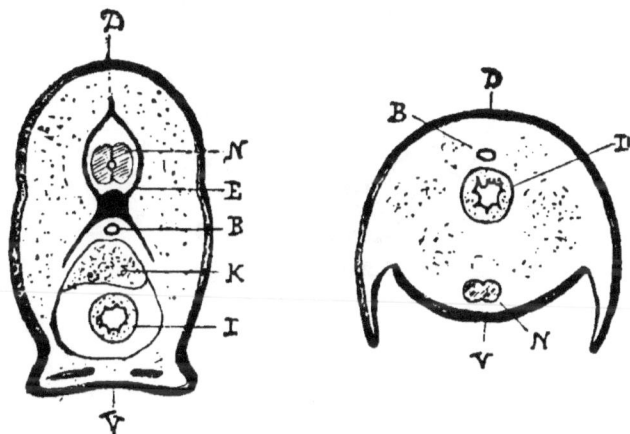

Fig. 119.—Diagram of transections through the abdominal regions of a catfish and a crayfish, to show the relative position of nervous system, N, and intestine, I.

D, dorsal; V, ventral surface; E, endoskeleton; B, aorta; K, kidney. •

3. The last peculiarity is especially met with in the **Crustacea,** one of the four Arthropod classes, and that to which the crayfish belongs. The other classes (**Insecta, Arachnida, Myriapoda**) embrace chiefly air-breathing Arthropods, whilst almost all Crustacea are aquatic, so that there is, on the whole, a marked difference between the respiratory organs of the Crustacea and those of the other classes.

Several species of crayfish or crawfish (old English crevish, Fr. écrevisse, Ger. Krebs) occur in Ontario; one of the commonest near Toronto is *Cambarus robustus*, Girard, (Fig. 120),

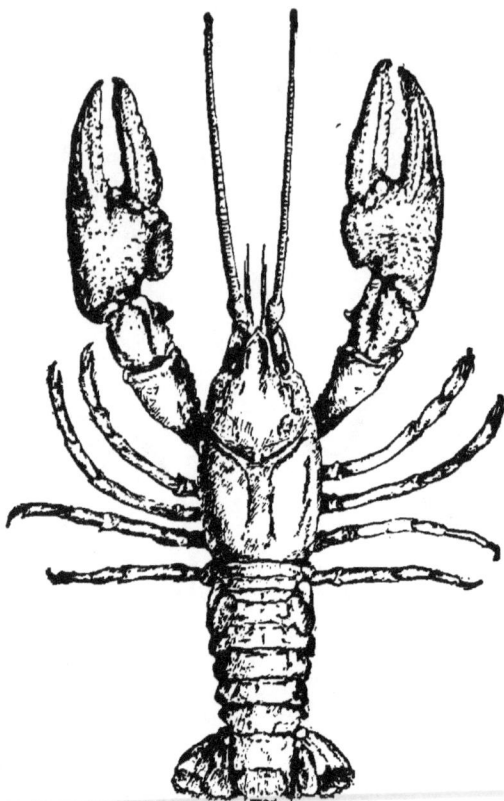

Fig. 120.—*Cambarus robustus.* (Girard).

but all are included in the genus **Cambarus**, while those on the Pacific slope of the Rocky Mountains, as well as the European crayfishes, belong to an allied genus, **Astacus**. This genus gives its name to the family Astacidæ, which includes the lobster (*Homarus americanus*), the marine representative of the crayfish. Any species will serve for making out the arthropodous characters already mentioned, as well as those peculiar to the Crustacea which follow.

4. Twenty segments, of which only the last (*telson*) is destitute of a pair of appendages, are invariably present in the group of Crustacea, to which the crayfish belongs, and these are

grouped in three regions, of which the head contains five, the thorax eight and the abdomen seven. On account of the extent to which a segment from one region differs from that from another, the segmentation is styled **heteronomous**, but the same fundamental plan of structure may be observed in all. The abdominal segments are independent, but the segments of the head and thorax are coalesced with each other into a **cephalo-thorax**, the fusion being more complete on the dorsal surface. Behind a line which marks off the cephalothorax into anterior and posterior regions, each side of the thorax is provided with a flap of skin which acts as a gill-cover, forming a cavity in which the gills, attached to the bases of the thoracic legs, are sheltered.

5. It will be convenient to study one of the hinder pairs of abdominal appendages first; they are biramous, consisting of a basal part, with two branches, internal (*endopodite*) and external (*exopodite*). Those of the sixth pair are modified with the telson into the caudal fin; while the first and second pairs are different in the two sexes.

Of the eight pairs of thoracic legs, the three foremost are turned forward as the foot-jaws (*maxillipedes*) to assist in securing food, while the five hindmost are the walking legs. Comparing these with the abdominal appendages, we find that although the endopodite is large in all, the exopodite is only present in the foot-jaws, while, with the exception of the eighth pair, all the thoracic appendages have in addition a membranous flap—the *epipodite*—concealed within the gill-chamber, and carrying, with the exception of the first, gill-filaments. There are thus six gills of this nature on each side; the other gills are attached to the soft membrane which connects the legs to the thorax, and there are eleven of these on each side, the third to the seventh appendages each carrying two, while the second has only one. The fourth pair of thoracic legs—the great claws—are adapted for prehension; they, like the fifth and sixth, are chelate, *i.e.*, the penultimate joint is prolonged so as to be opposed to the terminal joint.

Of the five pairs of head-appendages, the two anterior (*antennulæ* and *antennæ*) are sensory, while the three posterior (the *mandibles* and the *maxillæ*) are related to the mouth-aperture as jaws. The second pair of maxillæ most closely resemble the foot-jaws, but the exopodite and epipodite of each are united into a spoon-shaped flap, which lies in the anterior narrow aperture of the gill-cavity, and, by its movements, creates a current of water, which flows outward through that aperture. In both pairs of maxillæ as well as in the mandibles, the endopodites are feeler-like (*palps*), while it is the basal segments which are flattened and approximated to the mouth-aperture, those of the mandibles alone being hardened for cutting. Neither exopodites nor epipodites are present in the mandibles or first pair of maxillæ. On the other hand, both of the foremost appendages are biramous, the exopodites of the antennæ being, however, mere scales, while those of the antennulæ are similar to the endopodites. On the basal joints of the antennæ and antennulæ, respectively, are to be seen the apertures of the green glands or kidneys and of the ears to be afterwards described.

6. Having inspected the outward form of the body, we must now glance at the various systems of organs. It will be observed that the chitinous cuticle remains soft where movements are necessary, and that it is most densely calcified where it meets with the greatest strain, as *e.g.*, in the chelæ and mandibles.

7. The muscles are formed of very plainly striped tissue; indeed the histology of this tissue can be more easily studied here than in the catfish (I, 8). The muscular bundles are attached to ingrowths of the exoskeleton, which can be seen very

well in the chelæ and elsewhere. Similar ingrowths protect the ventral nervous system.

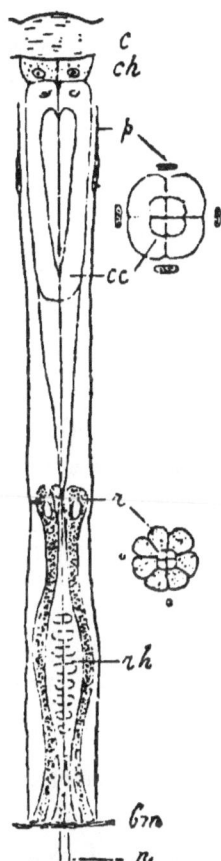

8. We distinguish in the nervous system, the brain and ventral nerve-cord, the latter composed of a chain of paired ganglia, connected by longitudinal commissures. Of such ganglia the last eleven segments in front of the telson have each one pair, but the ganglia of the five segments in front of these have coalesced into an infracœsophageal ganglionic mass. This is united to the brain, or supra-œsophageal ganglia, by commissures which lie at either side of the œsophagus. The brain supplies nerves to the eyes and antennæ. All of the nerves in the crayfish, as well as other Invertebrates, are of the non-medullated type.

9. An examination of the sense-organs shows that they differ both in position and structure from those of Vertebrates. The eyes are elevated in this order of Crustacea (*Podophthalmata*) on movable stalks, and they are of the compound type so characteristic of most Arthropods (Fig. 121).

Fig. 121.—Diagram of an ommatidium from the eye of the Crayfish ; c, cuticular facet formed by underlying hypodermal cells ch ; p, pigment cells surrounding the retinophoral cells and cc, the crystalline cone; r. retinulæ ; rh, rhabdome ; bm, basement membrane ; n, nerve.

The stalk is partly occupied by muscles, but chiefly by ganglionic expansions of the optic nerve, from the outermost of which the nerve-fibres pass off through a basement membrane to end in the modified epidermal cells which constitute the eyes. These cells are disposed in three zones, the outermost of which secrete the cuticular facets of the eye ; each facet corresponds to an element (ommatidium) of the compound eye, and is formed by two cells of the outermost zone; underneath these are the four retinophoral cells, (surrounded by four pigment cells), which secrete the crystalline cone, and this is prolonged inwards into a tube (formed by seven cells of the innermost zone—the retinulæ), of which it forms a spindle-shaped core—the rhabdome. The nerve fibre to each ommatidium occupies the axis of the rhabdome and of the crystalline cone ; the cones, therefore, constitute the sensitive elements of the eye, like the rods and cones of the Vertebrates.

Auditory sacs are present in the basal joints of the antennulæ. They contain foreign particles, which play the part of otoliths (I, 46), and the sensitive elements are stiff hairs in which nerve-fibres terminate. Both pairs of feelers obviously act as tactile organs, but peculiar setæ on the outer branches of the antennulæ have been interpreted as olfactory in function.

When the various jaws have been removed, the mouth is exposed, bounded in front and behind by unpaired chitinous outgrowths, the labrum and metastoma. The chitinous cuticle is continued into the spacious stomach, where it forms numerous calcified teeth, of use in comminuting the food. Digestive juices are furnished by the so-called liver, a bulky tubular gland which lies above and behind the stomach, and which opens into the mid-gut, the only part of the intestine destitute of chitin. Behind this is the straight rectum, the lining of which becomes continuous with the cuticle at the anus.

10. In comparison with the Vertebrates, the Arthropods have a less complete blood-vascular system, for, during part of the circulation, the blood flows in interspaces instead of closed capillary vessels. These are, however, partly represented in the crayfish, and the heart, as well as the arteries and veins, is well developed (Fig. 122). The blood is driven out to the whole system through the arteries, and is returned by venous sinuses through the gills to the pericardial sac.

11. The female crayfish may be found in spring with eggs attached to the abdominal appendages, to which the young adhere until they have attained the form of the adult. In the

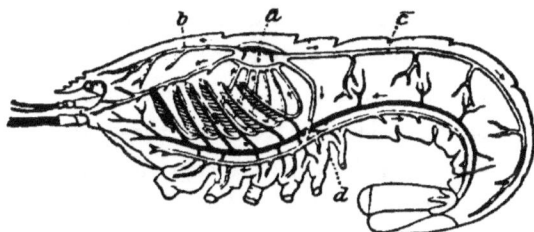

Fig. 122.—Diagram of circulation of Crayfish—*a*, heart in pericardium ; *b, c, d*, anterior, posterior, and ventral arteries.

lobster, as well as most other Crustacea, the young are freed from the egg when they have attained three pairs of legs (Nauplius-phase) ; they only arrive at the adult form after a series of moults, and there is generally a complicated metamorphosis.

12. If we except a species of prawn (*Palæmonetes*) and another of Opossum-shrimp (*Mysis*) found in the Upper Lakes, the Podophthalmata are exclusively marine forms, including on the one hand the various kinds of shrimps and prawns, which resemble the crayfish in the long tail (*Macrura*), and on the other, the crabs (*Brachyura*), where the short tail is tucked up under the cephalothorax. An intermediate group is formed by the hermit-crabs (*Paguridæ*), in which the cuticle of the tail-segments never becomes calcified, and the creatures resort to empty univalve shells for protection. An allied East Indian genus, the cocoanut crab (*Birgus latro*), lives in holes in the earth, and, instead of depending on its gills for respiration, uses the wall of the gill-cavity as a lung. This is an instance of what is termed "change of function," a principle which must be borne in mind, in comparing the structure of animals which are nearly allied in form, but different in habits.

13. Two other orders of Crustacea, which resemble the crayfish in the number of the segments and the appendages, have fresh-water representatives which are very common, although the majority of both are marine. These are the **Isopoda** and the **Amphipoda**; but in both, only one of the eight pairs of thoracic appendages is turned forwards toward the mouth. The Isopods have the body depressed, while in the Amphipods it is compressed. A familiar example of the former is the water-slater, *Asellus communis*. (Fig. 123). It will be observed

that the four hindmost abdominal segments are coalesced above into a shield, from beneath which the last pair of legs project. The three pairs of legs in front of these serve for respiration, and the eggs are carried in the female on the under surface of the thoracic segments. Terrestrial Isopods, like the common Wood-louse (*Oniscus*) and its allies, exhibit an interesting adaptation for breathing air; one of the pairs of abdominal legs being traversed by tubes which have the same function as the tracheæ of insects.

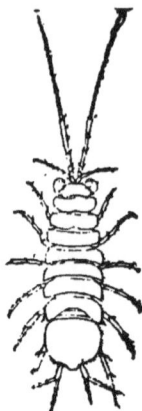

Fig. 123 –*Asellus communis.* × 2.

Among the marine Isopods several are temporary parasites adhering to the surface of fish; others are permanent parasites, which live in the gill- or body-cavity of other Crustacea, and which consequently loose much of their resemblance to the free Isopods.

Of the fresh-water Amphipods species of a genus *Gammarus* (Fig. 124), are everywhere to be met with. The gills are on the thoracic legs, the abdominal legs being partly for swimming, and partly for leaping. Species of an allied genus, *Pontiporeia*, occur in the Great Lakes; they are inter-

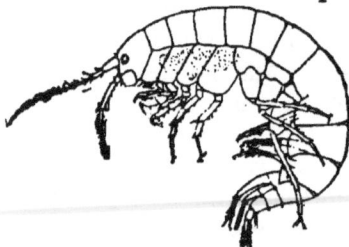

Fig. 124.—*Gammarus* sp. × 3.

esting, like Mysis, because the other species are chiefly marine.

14. The lower orders (separated in a sub-class **Entomostraca,** from the foregoing **Malacostraca,**) exhibit by no means the same constancy in the number of segments which we meet with in the higher, nearly allied forms often presenting considerable differences in this respect. All of the orders except one—the **Cirripedia**— have fresh-water representatives, which are for the most part inconspicuous, often microscopic, creatures.

The most primitive forms, as well as the largest we have to mention, belong to the **Phyllopoda,** a group in which all the segments behind the head bear flat leaf-like swimming-legs.

A common form in spring pools is *Branchipus vernalis*

Fig. 125.—*Branchipus vernalis*, swimming on its back.   × 3.

(Fig. 125), with eleven pairs of such legs ; an allied genus, *Artemia*, is very common in salt lakes. Other genera are protected by a shell, which may be horse-shoe shaped as in *Apus*, or formed of two valves as in *Estheria*. Such shells are also found in another group of Phyllopods—the **Cladocera** or water-fleas—(Fig. 126), in which there are only five pairs of legs, but which

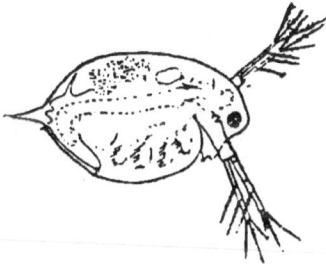

Fig. 126.—*Daphnia pulex*. × 18.

Fig. 127.—*Cypris candida*. × 16.

are otherwise marked by the large second pair of antennæ taking on a locomotive function. Another order, **Ostracoda**, includes forms with a still shorter post-cephalic region, for only two pairs of legs are to be found behind the jaws (Fig. 127). The **Copepoda**, however, have a much longer post-cephalic region than this, there being five thoracic segments, the first of which is coalesced with the head, and five abdominal segments terminating in a furca. The latter are footless, but the thoracic segments bear biramous swimming feet, and the head-segments the usual appendages, although the second pair of maxillæ separate on each side into two independent so-called foot-jaws, which may undergo curious alterations. Many of the Copepoda live a semi-parasitic or parasitic life ; in the free forms (Fig. 128), the jaws are formed for biting, but in the parasitic forms, the parts of the mouth are more or less converted for sucking or adhesion (Figs. 129 and 130), in which the posterior antennæ may assist. As a rule the parasitic Copepoda do not appear **to injure** much the creatures they attack, but one form, *Argulus*, which attains the length of quarter of an inch, is found to kill the whitefish in lakes in the North-West in immense numbers.

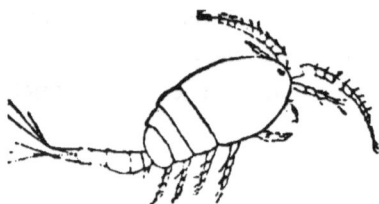

Fig. 128.—*Cyclops* sp.   × 12.

Fig. 129.-*Ergasilus* with egg-sacs from gills of sunfish × 10

Fig. 130.—*Ach-theres* from gills of catfish   × 6.

This form attaches itself by the anterior foot-jaws, which are modified into suckers, but it is the piercing and sucking mouth which injures the fish.

The remaining order — the **Cirripedia**—has only marine forms, which pass through an active larval phase, but eventually attach themselves by their heads and secrete a complicated shell (Figs. 131 and 132). The antennæ are rudimentary, but three pairs of jaws are present, and behind these, six pairs of biramous feet, which by their movements bring food particles to the mouth. These are, however, absent in certain parasitic forms.

Fig. 131.—Shell of *Balanus hameri*.

Fig. 132.—*Lepas anatifera.*

15. On the Atlantic coast, from Nova Scotia southwards, there occurs a very remarkable animal called, on account of its shape, the horse-shoe crab, *Limulus polyphemus* (Fig. 133), the body of which is divisible into three regions — cephalothorax, abdomen and caudal spine. The first of these bears six pairs of leg-like appendages, chiefly chelate, on either side of the mouth, possibly equivalent to the first six pairs of the Crustacean. The abdominal appendages are present in five pairs, the outer branches of which are beset with gill-leaflets. Limulus passes through a "Trilobite-phase" (Fig. 134), in its development, so called on account of its resemblance to the singular fossil Arthropods, which were so abundant during the Palæozoic period. (Fig. 135).

Fig. 133.
Limulus polyphemus. ⅓.

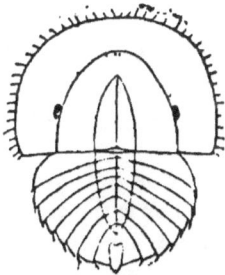

Fig. 134.-Trilobite-phase of Limulus. (After Kingsley).

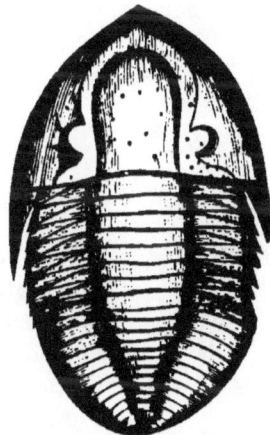

Fig. 135.—*Asaphhus Canadensis.* Chapman. Utica Formation.

16. Both the Trilobites and Limulus have generally been considered as Crustacea, but many points in the structure and

14

development of the latter seem to point to its being more closely
allied to the **Arachnida.** This resemblance is strongest to the
Scorpions (Fig. 136), a group of Arach-
nida confined to the warmer zones of
the Old and New Worlds. In these
the appendages of the cephalothorax
are similar to those of Limulus, the
first two pairs (**chelicerae** and **pedipalpi**)
acting as jaws and prehensile arms,
while the others are walking legs.
Respiration is performed by four pairs
of "lungs," which are cavities on the
third to the sixth abdominal segments,
containing leaflets that recall the gills

Fig. 136.—*Scorpio europæus,*
with its combs and ocelli.

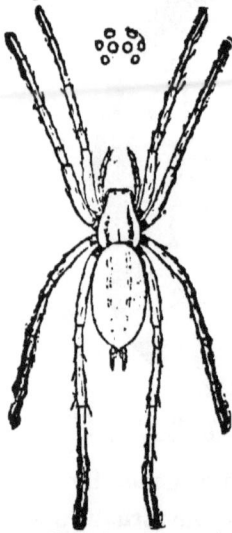

of Limulus, and opening by slits on the ventral surface.
Development shows that these lungs arise as infoldings at the
bases of appendages, and that they are homologous with the
gills of the horse-shoe crab. The abdomen differs from that
of the crayfish, in being differentiated into
two regions, a preabdomen of seven, and
a postabdomen of six segments, the last of
which terminates in a curved claw, perforated by the duct of a poison-gland.

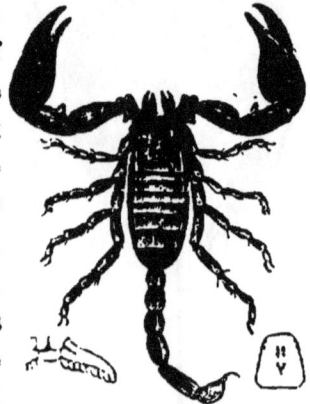

The little book-scorpions (*Chelifer*) have no
poison-gland in the tail, nor is the abdomen subdivided into two regions ; they belong to an independent sub-order, the *Pseudoscorpionina,* as
do the daddy-long-legs (*Phalangina*), with their
short abdomen and long walking legs. Both of
these groups feed on minute insects and mites ;
with the Scorpions they form the order **Arthro-
gastra.**

17. Of the various orders of Arachnida,
the spiders (**Araneina**) and mites (**Acarina**)

Fig. 137—*Agalena nœvia,*
with the ocelli
(After Emerton).

are the most important. Both have the two pairs of mouth-appendages and the four pairs of walking legs, but the form of the body is very different in the two groups, on account of the separation of the abdomen in the spiders proper, by a slender stalk, and the presence at its extremity of the spin-nerets (Fig. 137).

Some of the chief structural peculiarities of the spiders may be gathered from Fig. 138. The two-jointed cheliceræ terminate in a powerful claw, perforated by the duct of a

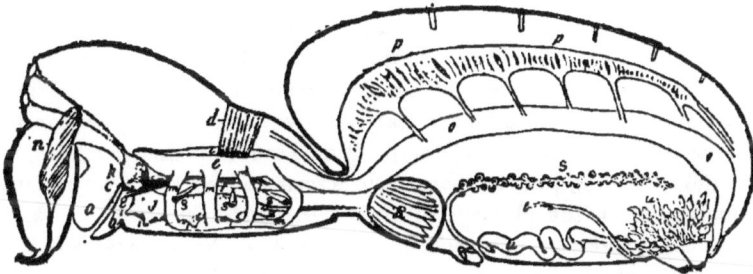

Fig. 138.—Diagrammatic section of a spider—*Epeira*. (After Emerton).

a, b, upper and lower lips ; c, œsophagus ; d, f, upper and lower muscles of the suck-ing stomach ; e, stomach ; g, ligaments attached to diaphragm under the stomach ; k, upper, j, lower, nerve-ganglion ; l, nerve to legs and palpi ; m, m, branches of stom-ach ; n, poison-gland ; o, intestine ; p, heart; r, lung ; s, ovary; t, trachea ; u, spinning glands.

poison-gland. Between the bases of the pedipalpi is the mouth, which leads by an œsophagus into a sucking stomach, dilatable by muscles, and provided with lateral cœca. The abdominal part of the intestine is provided with a liver, and with Malphigian tubes (slender cœca arising from the hinder end of the intestine in air-breathing Arthropods, and discharging the function of kidneys). The heart is elongated like that of the scorpion and of the lower Crustacea, but the nervous cord is concentrated into the thorax. Above the œsophagus is the brain, which sends nerves to the simple (not facetted) eyes, the arrangement of which on the head is of great use to systematists. The numerous lungs of the scorpion are only represented here by two air-sacs (four in the trap-door spiders), while, in addition, a pair of branching air-tubes, such as are universally present in the insects, open further back near the spinnerets. These are three pairs of projections, through short tubes on the ends of which the spinning glands open. The secretion furnished by these glands hardens on exposure to the air, and the threads so formed are guided by the hind legs into the characteristic webs, which serve as dwellings, or as traps for the prey, or even for flight.

18. In the Mites, on the other hand, the abdomen and cephalothorax are coalesced and unsegmented, while the mouth-appendages are frequently much modified by the adoption of a parasitic mode of life. Some of the Mites are parasitic on insects, during a larval stage, in which they have only six legs, afterwards seeking their food on plants (*Trombidium*). Some are aquatic forms, which may live free (*Hydrachna*), or parasitically on fresh-water mussels (*Atax*). Others are temporary parasites, like the ticks (*Ixodes*), but there are various forms which live a permanently parasitic life in the plumage of birds (*Dermaleichus*), or in the skin of mammals (*Sarcoptes, Demodex*). Finally, the cheese-mites and their allies (*Tyroglyphus*) have their mouth-parts adapted to the easy mode in which they secure their food. Plants are not exempt from the attacks of mites, for the species of one group (*Phytoptus*), in which the two hinder pairs of legs are rudimentary, make minute galls in the leaves of various plants.

The effect of the adoption of a parasitic mode of life is best seen in the genus *Pentastomum*, a form destitute of appendages, except for two pairs of hooks near the mouth, which lives in the nasal cavities of Carnivora.

With the exception of the larger mites and ticks, all of the above forms have no special respiratory organs, and this is the case also with the bear-animalcules (*Tardigrada*), a group of microscopic creatures living in moss, and feeding on minute larvæ or Rotifers. Like the latter, they may be desiccated and revived by moisture. They are associated with the Arachnida on account of the number of appendages, but the fourth pair of legs occupies the hinder end of the body.

19. Apart from such minute air-breathing Arthropods as are referred to in the above paragraphs, all have respiratory organs, consisting (with the exception of the lungs of the scorpions and spiders) of branched air-tubes, communicating with the outside by " stigmata," and introducing air into all the tissues of the

body. The walls of these tubes are delicate, but they are prevented from collapsing by the presence of a strengthening, spiral, chitinous fibre, just as, in the higher Vertebrates, the windpipe is by its cartilages. They are, therefore, called **tracheæ,** and the air-breathing Arthropods are hence frequently spoken of as the "tracheate" in contradistinction to the "branchiate" Arthropods. Two groups of the tracheate Arthropods remain for us to discuss—the **Insecta** and the **Myriapoda.** Although the latter contains the most primitive forms, yet some knowledge of a very accessible member of the former class—the red-legged grasshopper—will serve to introduce to both.

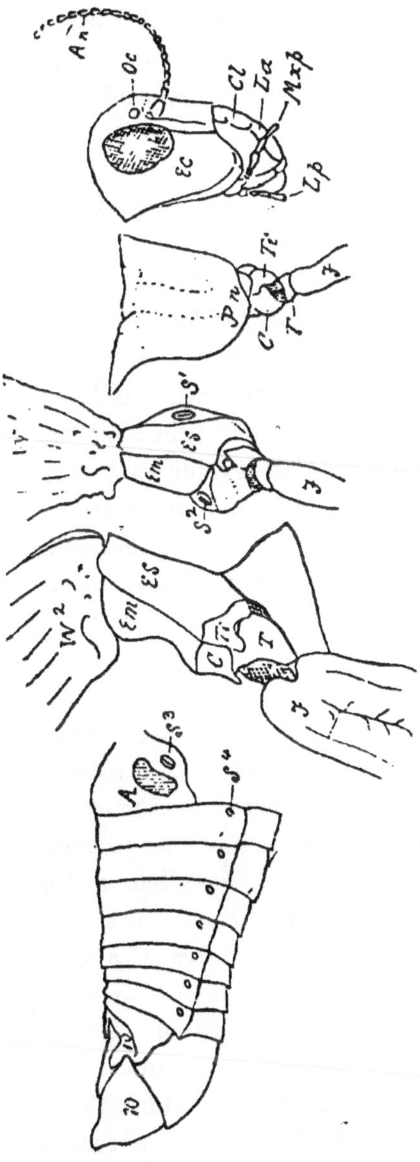

20. Among several species of locusts which are abundant in the fields in the fall, there is none easier to obtain than the species above

Fig. 139.—Segments of a Grasshopper. (After Kingsley). An, antennæ; Oc, ocellus in front of compound eye; Ec, epicranium; Cl, clypeus; La, labrum; Mxp, maxillary palp; Lp, labial palp; Pn, pronotum; C, coxa; T, Trochanter; Tl, Trochantine; F, femur; W¹ W² wings; Em, epimerum; Es, episternum; S² S² the thoracic stigmata; S³ S⁴ the first two abdominal stigmata; A, the ear; 10, the notum, (behind it the cercus); and 10, sternum of the tenth segment.

named—*Caloptenus femur-rubrum.* Comparing it with the crayfish, we find that there are conspicuous differences both as to the number and grouping of the segments, and as to the number and nature of the appendages. Instead of a cephalo-thorax and abdomen, we find a head composed of four segments (the number only to be arrived at from the appendages), a thorax consisting of three free segments (pro- meso- and meta-thorax), each of which bears a pair of walking legs, and the two hindmost, each a pair of wings ; and finally, an abdomen of ten segments without obvious appendages (Fig. 139).

21. In the abdomen, the exoskeleton of each segment is divi-sible into a *sternum* below, a *tergum* above, and a lateral piece on each side—the *pleurum*—coalesced with the tergum, and only indicated by the stigmata. In the meta- and meso-thorax (but not in the prothorax) a further differentiation is associated with the attachment of the wings, for each tergum or *notum* is subdivided into an anterior *scutum* and a posterior *scutellum,* while the independent pleurum is subdivided on each side into an anterior *epimerum* and a posterior *episternum.* Only the sterna of the head-segments can be recognized, for the dorsal part of the exoskeleton of the head (*epicranium*) be-longs solely to the first segment.

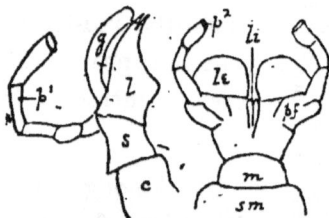

Fig. 140.—Maxilla and labium of Caloptenus. (After Packard). c, cardo ; s, stipes ; l, lacinia ; g, galea; p¹, maxillary palp ; sm, submentum; m, mentum ; pf, palpifer; le and li, external and internal lobes ; p², la-bial palp.

The thoracic legs are formed of the femur, tibia, and three-jointed "tarsus,"—these names must not be supposed to indicate any homology with the parts so-called in Verte-brates—articulated to the thorax by the "trochanter," "coxa" and "tro-chantine," but the head-appendages are more complicated. They em-brace the filiform antennæ, the strong cutting mandibles, the max-illæ and the **labium,** which is formed of a second pair of maxillæ, coalesced in the middle line (Fig. 140). Certain unpaired struc-

tures, like the labrum and metastoma of the crayfish, are represented here also, for above the mandibles there is an unpaired
labrum articulated to the epicranium by an intermediate *clypeus*, and projecting into the mouth-cavity as the *epipharynx*,
while a *hypopharynx* is found opposite in the floor of the
mouth ; both of these are covered with stiff hairs.

Although the abdomen has no obvious appendages yet the
blades of the ovipositor and the cerci (more conspicuous in the
cockroach) are, in reality, appendages of the eight, ninth, and
tenth segments ; traces of an eleventh segment are also present.

22. Of very different nature from the appendages, are the
wings : these are to be regarded as outgrowths from the notum
of the two hinder thoracic segments, which have become hinged
to the thorax, and penetrated by vascular and respiratory
organs. In this genus the anterior wings are less of use in
flight than the posterior, and serve partly as wing-covers (*elytra*).

23. In the locust, the nervous system is related to the segmentation in a way somewhat similar to what we found in the
crayfish, but it is not so concentrated as in the spider. The
brain supplies the eyes, ocelli and antennæ; the infraœsophageal
ganglion, the mouth-parts; the three thoracic segments have each
their own ganglion (the last of which supplies the ears) ; but
there are only five abdominal ganglia situated in the third,
fifth, seventh, eighth and ninth segments respectively (Fig.
141). The intestines have a special ganglion united with the
brain by a visceral nerve.

Little definite is known with regard to the senses of the locust ;
the antennæ and palps of the jaws have undoubtedly a tactile
function, but it is likely that some parts of them may be employed
to detect odours and tastes as well. The compound eyes have
a similar structure to those of the crayfish, but there are in
addition three ocelli (comparable to a single ommatidium of a
compound eye), one of which is situated between the bases of
the antennæ, the other two, higher up on the front of the head.

Locusts are capable of producing sounds by rubbing the hind legs against the wing-covers, and they have also organs fitted for perceiving sounds. These are situated on the first abdominal segment, and consist of a vesicular auditory sac, suspended on a stretched tympanic membrane.

24. Fig. 141 indicates the chief parts of the intestinal system. The œsophagus, on ascending from the mouth-cavity (into which salivary glands open), dilates into a crop, the lining of which is furnished with hairs regularly arranged. At the junction of the crop with the stomach are several cœca, which serve to increase the intestinal surface, while at the junction of the stomach with the intestine proper, the Malphigian or urinary tubes are situated. The heart is an elongated vessel occupying the first seven abdominal segments, and the respiratory organs are a complicated system of air-tubes, opening by ten pairs of apertures (stigmata or spiracles) to the outside. Two of these are thoracic, being situated in front of and behind the mesothorax, while the other eight are on the anterior eight abdominal segments. The spiracles communicate directly with two longitudinal lateral air-tubes; these give off the smaller tracheæ to the tissues, but the spir

Fig. 141—Diagrammatic longitudinal section of Caloptenus. (After Burgess).

B, brain ; O¹ O², ocelli ; C, crop ; Cœ, cœca ; H, heart ; Ov, ovary ; S, stomach ; Sa, salivary glands ; I, Infracœsophageal ganglion ; In, intestine, (between In and S, the Malphigian tubes come off) ; Od, oviduct ; Op, ovipositor ; E, egg guide.

acles are also very directly related to a series of large air-sacs, which buoy up the locusts in their flight.

25. There is a conspicuous difference between the end of the abdomen in male and female specimens ; in the latter (Fig. 141), the ovipositor serves to drill the holes in the ground in which the eggs are laid, surrounded by a stiff secretion furnished by special glands. Oviposition occurs in the fall, and development begins at once, but is checked by winter, so that the young larvæ only escape from the eggs in the spring. Apart from the circumstance that they are destitute of wings, they resemble the parent in form ; the complete resemblance is attained during a series of moults, after each of which the body becomes larger and the rudimentary wings more evident. No complete resting-stage occurs such as the "chrysalis" of the butterfly, but the insect is said to be in the " pupa" stage, before the last moult, which converts it into the adult (imago) stage : the locust and its allies are consequently said to develop without metamorphosis.

26. According to a recent computation, the number of species of living animals described, amounts to some 272,000 ; of these, 200,000 belong to the class of the Insecta, and are consequently constructed upon substantially the same plan as the locust described above. Although only some 6,000 of these occur in Canada, yet there is such a wealth of form, and such differences of habit within the limits of this single class, that it will be impossible to do more here than indicate the chief modifications of the insect type, which characterize the various orders.

Most of these are more specialised than the type described, so it may be as well to glance in the first place at the more primitive forms. Such, like the cockroaches (*Blattidæ*), and earwigs (*Forficulidæ*), are to be found within the order (Orthoptera) and sub-order to which Caloptenus belongs. The order receives its name from the position which the wings assume in rest in the family (*Acrydidæ*), in which it is placed, and to which the carnivorous locusts and crickets are nearly allied; but the wings may be entirely absent, as in the singular walking-stick insects (*Diapheromera femorata*), or only partly developed, as in some of the cockroaches and earwigs.

The family *Phasmidæ* contains some of the most striking cases of protective resemblance to the environment, for the members may resemble dried twigs, or even leaves of the trees on which they live, as in the case of the East Indian winged *Phyllium*.

27. From the earwigs we are led to a group of insects characterized by the entire absence of wings, and the presence of caudal appendages, equivalent to the cerci of the locust and cockroach, and to the forceps of the earwigs. These are the spring-tails (**Thysanura**), inconspicuous on account of their size (Fig. 142), but interesting to the zoologist as the most lowly organized insects, some of them (Fig. 143) even having rudimentary legs on the abdomen, and thus resembling certain Myriapods (Fig. 144).

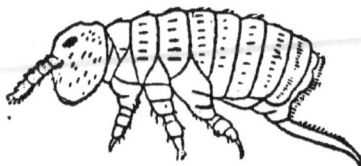

Fig. 142.—Podura.

Fig. 143—Campodea.

Fig. 144.—Scolopendrella.

28. The characters of this class, the **Myriapoda**, may be therefore briefly examined before proceeding to the higher Insecta. It is a small group of less than a thousand species, in which the numerous segments of the body may each bear one or two pairs of appendages, but are never grouped, as they are

in the Insecta, into thorax and abdomen. There are no wings, and the maxillæ may only be present in one pair. Most of the Myriapods fall into two orders, the **Chilopoda** and **Chilognatha**, the former including carnivorous forms (Centipedes, Fig. 145), the latter, forms which live in decaying vegetable material (Millipedes and galleyworms). The parts of the mouth are adapted to their habits, for in the Chilopoda the first pair of legs end in powerful claws perforated by the duct of a poison-gland, and are turned forwards to supplement the three pairs of jaws. In the Chilognatha, on the other hand, there is only one pair of maxillæ below the mandibles, and they are united to form a labium. The two groups further differ in that the Centipedes have only one pair of legs to each segment, the Millipedes two; and that, while the Centipedes resemble the insects in the position of the opening of the oviduct, this is near the head in the Millipedes. It will be apparent from what follows that these are much more important structural peculiarities than we find separating the orders of Insecta from each other.

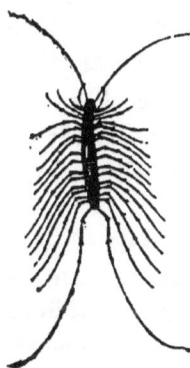

Fig. 145.—Scutigera.

Certain tropical worm-like forms (*Peripatus*) which have the habits of Millipedes, but whose segments bear unjointed appendages terminating in hooks, are of interest as being intermediate between the Vermes and the lower Arthropoda. A separate class (**Protracheata**) has been formed for their reception.

29. Returning to the locust and its allies, which are described as the Orthoptera proper (*O. genuina*), we must now proceed toward the higher orders of insects, glancing, in the first place, at certain forms associated by naturalists with the Orthoptera, on account of the structure of the mouth-parts and the absence of a metamorphosis, but differing from them in that both pairs of wings are alike. The wings resemble those of the nerve-

212 HIGH SCHOOL ZOOLOGY.

winged insects (Neuroptera), and, to distinguish them from
these, the forms referred to are called *Pseudo-neuroptera.*
Belonging here are the dragon-flies (*Libellulidæ*), May-flies
(*Ephemeridæ*), stone-flies (*Perlidæ*), all of which have aquatic
larvæ (into the tracheæ of which air is absorbed through
peculiar expansions of the body-wall known as tracheal gills),
but there are also forms with terrestrial larvæ, such as the
Psocidæ (very small insects which live like plant-lice chiefly on
hardwood trees, and often attract attention by the woolly-looking
masses which they form).

Allied to these are the tropical Termites, often called "white ants,"
because they live a social life in colonies and build nests. An African
species—*Termes bellicosus*—builds towers 12 or 15 feet high ; in addition
to the males and females, the inhabitants are partly wingless neuters,
most of which undertake the work, but some the defence of the colony,
and are therefore called workers and soldiers.

Occupying an intermediate place between the Orthoptera and the next
order, is the family *Thrypsidæ*, including minute insects which have the
parts of the mouth adapted for sucking vegetable juices. They often
attack cultivated plants in great numbers, causing destruction, *e.g.*, of
the hay and onion crops. The wings of the adults are margined by long
delicate hairs.

30. Like the Orthoptera, the **Hemiptera** are insects with an incomplete metamorphosis, but the parts of the mouth are generally modified for sucking, the labium being converted into a
grooved and jointed proboscis (generally folded back underneath
the thorax), in which the mandibles and the maxillæ lie in the
form of slender stylets (Fig. 146, 2).

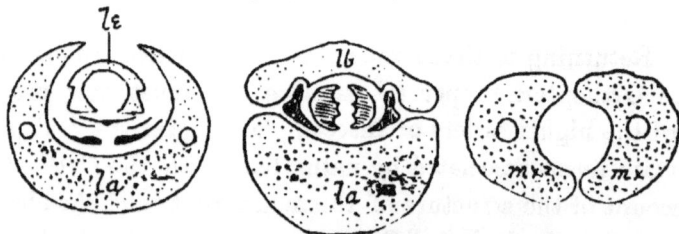

Fig. 146, 1, 2, 3.—Diagram of transections of the proboscis of dipterous, hemip-
terous, and lepidopterous insects. (After Dimmock, Graber, and Muhr, respectively).
le, labrum and epipharynx; *lb*, labrum, *la*, labium , (between the two are the
mandibles and maxillæ, and in 1, the hypopharynx); *mx*, maxillæ.

Palps are absent, except in some low wingless forms—the *Mallophaga*—which live on the young hairs and downs of mammals and birds, and have, consequently, biting mouth parts. They, with the *Pediculidæ*, which live by sucking the blood of animals, form the sub-order *Aptera*. But the absence of wings is not confined to this sub-order, some of the plant-parasites (sub-order *Phytophthires*) being also wingless. This sub-order includes the plant-lice, *Aphidæ*, and the scale-insects, which are so harmful to plants (especially those cultivated in the house)* the juice of which they suck by means of the proboscis.

Among the better known forms of these parasites are the *Phylloxera vastatrix*, which was carried from this continent to Europe, and has there done enormous damage to the vineyards, chiefly by forming galls on the rootlets of the vines, and the cochineal insect (*Coccus cacti*), a scale-insect which lives on the prickly-pear cactus in Mexico, and which is the source of carmine, one of the most valuable dyes of commerce.

The bulk of the Hemiptera belong to the two remaining sub-orders, the *Homoptera*, in which the wings are alike, and the *Heteroptera*, in which the anterior are partly converted into elytra. To the former belong the musical *Cicadidæ*, the males of which have a vocal apparatus on the under surface of the abdomen, and the *Cicadellidæ*, which are much smaller forms but include many more species. To the latter, belong the water- and land-bugs, *Hydrocores* and *Geocores*. In accordance with their aquatic life, the Hydrocores have one or more pairs of legs modified for swimming, their habitual mode of locomotion during the day, but their hind wings enable them to fly, which they do chiefly at night. All of them are predaceous forms, sucking the blood of fishes, Ephemerid larvæ, etc.; they are capable of inflicting a sting by means of their proboscis. The body may be elongated (*Ranatra*), or flattened (*Belostoma*), or keeled for swimming on the back (*Notonecta*). The Geocores, however, include many more species, partly living on animal, partly on vegetable juices. Some have extremely long legs, by

214 HIGH SCHOOL ZOOLOGY.

means of which they run over the surface of water (*Gerris*) ;
others have a flat, depressed body, with short legs, like the
bed-bug (*Acanthia lectularia*), while among the phytophagous
forms, to which some destructive species like the chinch-bug
(*Rhyparochromus leucopterus*), and the squash-bug (*Coreus
tristis*) belong, a great variety of form exists.

31. All the Insecta mentioned above are spoken of as ame-
tabolic forms (*Ametabola*), on account of the fact that they do
not undergo a metamorphosis. Those on the other hand now
to be dealt with are metabolic (*Metabola*).

At first sight the **Neuroptera**, on account of the wings, seem
to be closely allied to the May-flies referred to above, but their
larvæ pass through a resting (pupa) stage, during which they
attain their adult form. The terrestrial larvæ of *Myrmeleon*
are called ant-lions, as they feed on ants, which they catch by
preparing sand-pits for them to roll into. They spin a cocoon,
in which they pass their pupa-stage. The aquatic larvæ of the
caddis-flies (*Phryganea*) live in cases, formed of sand or bits of
twigs, in which they afterwards pass the pupa-stage. This
group (*Trichoptera*) is an interesting one, because it leads to
the Diptera and Lepidoptera, both on account of the fact that
the anterior wings are hairy, and because the mouth-parts approach the structure met with in these orders. The latter is
true also of the genus *Panorpa*, which is further remarkable for
having caterpillar-like larvæ.

32. From the carnivorous terrestrial larvæ of the Neuroptera
we pass naturally to the carnivorous forms of beetles—**Coelop-
tera**,—which are marked by a more complete conversion of the
anterior wings into elytra than we have yet met with, and by
a greater resemblance of the mouth-parts to those of the Orthoptera, than exists in other insects. More than a third of the known
species of insects belong to this order ; it may be gathered, therefore, that within comparatively narrow limits of structural
modification there is a surprising wealth of form, associated

with adaptation to habits of life characteristic for each species. Thus we have the carnivorous *Carabidæ*, running forms often destitute of the hind wings, the allied aquatic water-beetles, *Dytiscidæ*, and scavenger forms like the *Sylphidæ* and *Staphylinidæ*, several of the latter being always found in ants' nests, and presenting curious instances of dependence upon their hosts. Destructive household-pests like the *Dermestidæ* (which attack furniture, carpets, museum specimens, etc.), are amongst the smallest of the order, while the *Scarabæidæ*, along with a number of familiar phytophagous forms like the May-beetle (*Lachnosterna*), include some tropical giants—*Dynastes*—which may attain a length of 5 or 6 inches. The fire-flies (*Lampyridæ*) are sufficiently marked by the luminous organs on the abdomen, the weevils (*Curculionidæ*) by the prolongation of the head into a sort of proboscis, the bark-beetles (*Bostrychidæ*) by the characteristic channels which they hollow out in trees. Several of the leaf-eating forms (*Chrysomelidæ*), like the potato-beetle (*Doryphora decemlineata*), are familiar, and the lady-birds (*Coccinellidæ*), which feed on plant-lice, attract attention as well by their form as by their colouration.

33. In the two next orders of insects, **Diptera** and **Lepidoptera**, the mouth-parts are formed for sucking, but this conversion is carried out in different ways in the two groups. In the former the labrum and the labium form an unjointed double tube, in which stylets formed of the mandibles, maxillæ, and hypopharynx are contained (Fig. 146, 1). In the Lepidoptera on the other hand, it is the maxillæ which by their apposition form the sucking proboscis (Fig. 146, 3); the maxillary palps are rarely well developed, while the labial palps are. The Diptera receive their name on account of the apparent absence of the posterior wings, which are converted into balancers—*halteres*; the Lepidoptera, theirs, from the presence of scales (which are generally coloured) on the wings.

To the Diptera belong the mosquitos (*Culicidœ*), black-flies (*Simulidœ*), and horse-flies (*Tabanidœ*), the females of which suck blood through wounds made by their piercing stylets. The larvæ are aquatic, or, as in the case of the horse-flies, they live in the earth. There are also forms like the Hessian-fly (*Cecidomyia destructor*), which are injurious to cultivated plants, eggs being deposited within the cellular tissue, and thus forming galls, in which the larvæ are developed. Again, there are the domestic flies (*Muscidœ*), in which the ends of the labium—(*labellœ*)—are converted into a rasping proboscis, which enables them to dispense with their piercing stylets. The larvæ (maggots) lead a parasitic or saprophytic life. Finally, the fleas (*Pulicidœ*) are distinguished by the absence of wings and by the serration of their mandibles, which adapts them better for their life of semi-parasitism.

Many differences of habit are likewise met with among the Lepidoptera. Small forms like the clothes-moths (*Tinea*) belong to the *Microlepidoptera*, which also include a host of forms destructive to vegetation, in one way or another, like the coddling-moth (*Carpocapsa*). Such is also the case among the larger forms, *Macrolepidoptera*, which include the butterflies (*Papilionidœ*), hawk-moths (*Sphingidœ*), silk-worm moths (*Bombycidœ*), and other families.

It is in this order that remarkable instances of protective resemblance to other animals (so-called mimicry) was first observed. The bee-moths (e. g., *Sesia thysbe* and others) receive their names from (and owe their freedom from attack to) their resemblance to various stinging wasps.

34. The most highly developed of all insects are undoubtedly the **Hymenoptera.** They exhibit this in the reduction of the number of the abdominal segments and in the concentration of the nervous system, as well as in the social life which characterizes the higher genera. It is possible to recognize in the parts of the mouth, all of those met with in the locust, but the characteristic "tongue" of the bee is formed by the fusion and prolongation of the inner lobes of the labium, while the external lobes so marked in the locust here only form the paraglossæ (Fig. 147).

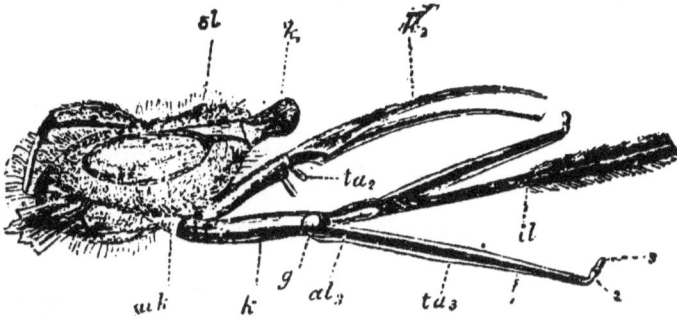

Fig. 147.—Part of the Head with the Mouth-parts of the Honey Bee.

Ol, labrum; k¹, mandible; k², maxillæ; ta², maxillary palps; uk, submentum; k, mentum; g, palpifer; al³, external lobe of the labium (paraglossæ); il, internal lobe, (tongue); ta³, labial palps.

Both pairs of wings are transparent, the anterior larger than the posterior, and united with them in flight. The lowest group, the saw-flies (*Tenthredinidæ*), have caterpillar-like larvæ which live on leaves, but are distinguishable from caterpillars by the greater number of abdominal feet. Most of the families, however, either provide for the larvæ, cells of wax, or of paper, or in the soil, and stock these with suitable food, which may consist of honey, or pollen and honey, or of paralyzed insects (*Sphegidæ*), or else they deposit their eggs in the bodies of other insects, or in plants, in such a way that the larvæ on hatching are surrounded with suitable food. In either case the larvæ are footless, and have rudimentary mouth-parts. The first condition is met with in bees (*Apidæ*), wasps (*Vespidæ*), and ants (*Formicidæ*), the females of which are furnished with a sting or modified ovipositor connecting with a poison-gland, the second in the *Ichneumonidæ*, which possess long ovipositors by which they deposit their eggs in other insects, and in the *Cynipidæ* (the gall flies), where the laying of the eggs by the short ovipositors, in the cellular tissue of plants, gives rise to characteristic diseased outgrowths—galls—, in the interior of which the insects live till they reach the adult condition.

15

CHAPTER VIII.

The Vermes and Mollusca.

1. In last chapter, reference was made to the prevailing heter-onomy of segmentation of the Arthropods, but the class of the Myriapods, and especially the genus Peripatus, were described as possessing a more worm-like form and homonomous segment-ation. The latter genus, indeed, is destitute of the jointed appendages almost universal in the Arthropods, and its locomo-tive organs rather suggest the unjointed stumps of the highest worms. These belong to the class **Annelida** (of the sub-kingdom **Vermes**), a class which is chiefly represented by marine forms, but of which the earthworm and leech may be selected as more accessible types.

2. The Vermes admit of no such sharp definition as do the Arthropods, for although bilateral symmetry is present in all, yet some forms are segmented, others unsegmented, and thus, the structure of the body may be extremely different in the various classes. The highest class, the Annelida, includes all the segmented forms, and those, consequently, nearest the Arthropods : a comparison, however, of an earthworm and a leech, on the one hand, with a crayfish on the other, will disclose the essential differences which exist between them.

3. One of the most notable of these is the absence of any exoskeleton such as that of the Arthropods ; a thin cuticle containing chitin represents it, and is formed by underlying epidermal (so-called hypodermal) cells, but many of these are

glandular in their character, and thus the skin is softer than in the Arthropods. The external segmentation does not always correspond to the internal; in the earthworm it does, partitions or septa being attached opposite the external furrows, which tend to divide off the cœlom into so many chambers as there are segments. In the leech, however, there are several furrows, which are merely skin deep, to each true segment marked off by the septa. The completeness of the internal segmentation, brought about by these septa, necessarily affects the various organs contained in the cœlom, thus the intestine, the blood-vessels and the nerve-cord all partake in it. It is not often that we observe any reduction in the number of repeated parts, (each segment, for example, having its own nerve-ganglion), and this is especially true of the excretory system, each segment having a pair of coiled tubes, segmental organs or **nephridia,** which open outwardly and also into the cœlom (Fig. 148). Ho-

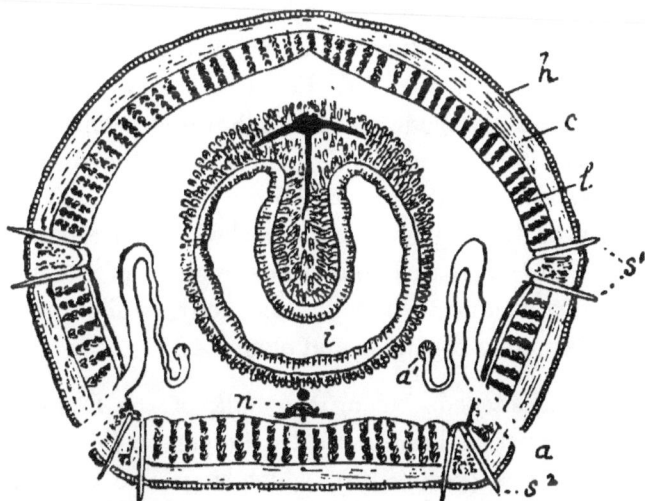

Fig. 148.—Diagram of transection of earthworm.

H, the hypodermis; *c*, the circular, *l*, the longitudinal muscular layers; *S*¹ *S*², the upper and lower pairs of bristles; *a*, external, *a*¹, cœlomic aperture of the nephridium (the external aperture is not exactly in the same plane as the setæ, nor is the internal aperture in the same plane as the external); *i*, the intestine with its roof infolded (typhlosole): it is coated with glandular epithelial cells; blood-vessels are represented in black, above and below the intestine, and around the nerve-cord—*n*.

mologues of these are to be recognized in the green gland of the
crayfish, and the coxal glands of various other primitive Arthropods, such as the scorpions and mites, which have a similar
position, but are not repeated in every segment as in the
Annelids.

The excretory organs of Annelids find a closer parallel in the kidneys
of the more primitive Vertebrates, which are also disposed segmentally;
the internal apertures may be retained, but the separate external apertures are replaced by a single collecting duct leading to the outside.

4. Instead of the elaborate muscles which are present in the
highest Arthropods, we find that the muscles and the skin
are closely united into a tube surrounding the cœlom. Immediately underneath the skin is a layer of circular fibres,
within that, one of longitudinal fibres, and both are penetrated
by radial fibres which extend from the skin inwards. Although
locomotion is always effected by the alternate contractions
and relaxations of this cutaneo-muscular tube, yet the precise
way in which it is carried out differs in the two subclasses of
Annelids—the **Chætopoda** and **Discophora**—to which the earthworm and the leech respectively belong. In the former, locomotion is assisted by bunches of strong bristles (setæ) attached
to the sides of the segments, and worked by special muscular
slips, while in the latter, one or more regions of the tube are
modified into suckers, which fix the body, while the muscular tube alternately contracts and elongates.

5. To the Chætopods belong the bulk of the Class, marine
forms with numerous bristles (Order **Polychæta**) fixed on short
projecting stumps (**parapodia**), which may also carry feelers,
gills, or protecting scales. The marine Annelids are either carnivorous in their habits, living a free life, and swimming or
creeping about the seashore, or sedentary forms, which burrow
in the sand (Fig. 149), or live in tubes of chitin or sand or lime,
which are constructed with the aid of secretions from the skin.

The order to which the earthworm belongs, however, (Oligochæta) chiefly includes fresh-water (limicolous) or terestrial (terricolous) worms, where the bristles are few in number, and lodged in setigerous follicles, there being no parapodia, nor appendages of the nature of feelers or gills. The Limicolæ are small forms living for the most part in the mud at the bottom of ponds or streams. Some of them, the *Naididæ*, are particularly remarkable on account of their reproducing themselves by budding, so that they are often found in chains, still attached to each other. They live chiefly on decaying vegetable

Fig. 149.—Marine lob-worm. (*Arenicola piscatorum*).

The bunches of setæ are more apparent in front of the gills than behind.

matter, but one species of *Chætogaster* lives a parasitic life in the lungs of various pond-snails. The earthworm (*Lumbricus terrestris*) is the most familiar of the Terricolæ ; its setæ are not conspicuous, but each segment carries eight, disposed in four groups. One region of the body is often swollen and noticeable, bearing the *clitellum*; it furnishes a cocoon in which the eggs are developed. The researches of Darwin proved the earthworm to be of the first importance in the loosening of the soil and the formation of mould. This is effected in the course of its burrowing, which it does partly by separating the particles of earth, partly by swallowing them. Although no special respiratory organs are present, yet the skin is traversed throughout by capillary vessels, which bring the blood close to the surface. The fluid portion of the blood (not the corpuscles, I, 60) contains hæmoglobin ; and its circulation through the skin, as well as through the rest of the body, is assured by heart-like loops which connect the principal longitudinal vessels. These are situated above and below the tubular intestine, and above and below and on each side of the nerve-cord.

6. The Discophora differ from the other Annelids in the entire absence of bristles, and in the presence of a sucker immediately below the posterior aperture of the intestine. Most forms also have the segment in front of the mouth (prestomium) converted into a sucker. In accordance with their habits of life, we find conspicuous structural differences in the Leeches, as compared with the other Annelids. They are sometimes temporary para-sites, like the medicinal leech in Europe (*Hirudo*) or its Ameri-can representative *Macrobdella*, living upon the juices of other animals, which they suck and store up in a sacculated intess tine, or carnivorous forms with a simpler intestine preying on the smaller Invertebrates (*Nephelis*). Both of these groups have jaws, which are used either for inflicting a wound be-fore sucking, or for comminuting their food. Other para-sitic forms are destitute of jaws, but have instead a protractile proboscis. Common examples are the fish-leech (*Piscicola* or *Ichthyobdella*), and the various species of *Clepsine*, large form-of which attack the pond-turtles, while smaller species prey on the pond-snails. Another interesting form, destitute of the anterior sucker, is the curious little *Branchiobdella*, several species of which are to be found on the gills, and on the head-parts, of the various species of crayfishes.

7. In contrast to the Annelids, the lower Classes of Vermes do not possess metameric segmentation. Certain marine forms—**Gephyrea**—have segmental organs, but the cœlom is undivided and the nerve-cord alike throughout. In the other Classes (**Rotifera, Nematelminthes,** and **Plathelminthes**), to which, for the most part, unfamiliar and inconspicuous forms belong, a

so-called water-vascular system—to be presently described—replaces the segmental organs.

8. The Rotifera or Wheel-Animalcules are microscopic aquatic creatures, round the mouth of which are disposed lobes bearing cilia, which, when in motion give the appearance of rotating wheels (Fig. 150). They serve to bring food to the mouth, and also for swimming. A longer or shorter tail, which is sometimes telescopic, assists in locomotion, and serves also, as it terminates in a pair of forceps, for temporary fixation. Some of the species are sedentary, living in tubes, either singly or in colonies. The intestine is absent in the males (which are not only more minute than the females, but also much fewer in number, and shorter-lived), but it is complete in the females, having near the mouth an expanded part containing a chitinous masticating apparatus, and, behind that, two lateral cœca. The water-vascular system consists of two convoluted tubes opening anteriorly into the cœlom and posteriorly into a contractile bladder, which communicates with the rectum. Instead of the elongated nerve-cord of the higher Annelids, there is here only a single ganglion situated above the œsophagus, whence are distributed nerve-fibres to the various parts, including the eyes and tactile organs. There are no blood-vessels, and no respiratory organs. The Rotifers like the Tardigrades have considerable power of resisting death by desiccation.

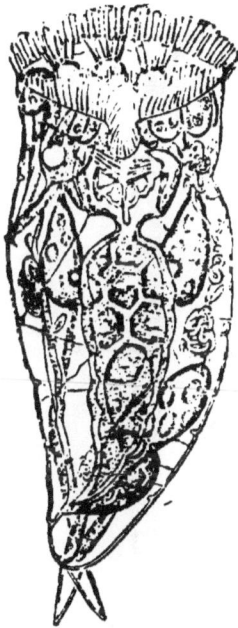

Fig. 150.—Female of Rotifer. (*Hydatina senta*).

9. The names of the remaining Classes, which include for the most part parasitic worms, are taken from the form of the body, which is cylindrical in the Nematelminthes but flattened in the Plathelminthes. According to the grade of parasitism, the intestinal system is more or less reduced, being absent in those most completely adapted for a parasitic mode of life. Organs of fixation suited to the nature and degree of the parasitism, are

also to be recognised. The development is often complicated
by a metamorphosis, and the creatures often pass through different stages of their parasitic life in different hosts, a phenomenon known as **heterœcism.**

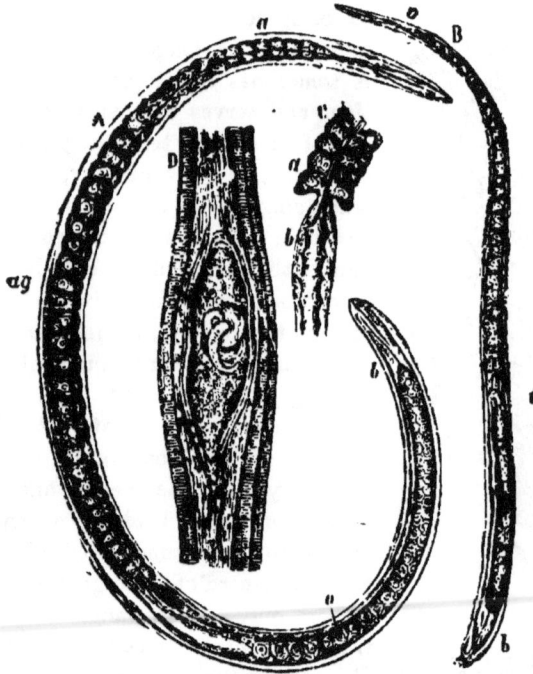

Fig. 151.—*Trichina spiralis.*

A, female; *B*, male; *C*, junction of œsophagus and intestine; *D*, encysted
Trichina-larva between the fibres of muscular tissue.

10. Two orders of Nematelminthes are recognised, the **Nematodes,**
which have generally a complete intestine, and the **Acanthocephali,** which
have none. To the former order there belong some free microscopic forms,
which live in decaying matter in water, also some vegetable parasites, like the wheat-worm (*Tylenchus*) and the beet-worm (*Heterodera*),
but the bulk of the order are parasites, either during a part of their
life (like the insect parasites *Gordius* and *Mermis*), or during the whole of
it. All groups of Vertebrates have special Nematode parasites, which live
in the skin, or the eyes (*Filaria*), or in the intestine (*Ascaris*), or the

muscles *(Trichina)*, or the respiratory organs *(Syngamus, Strongylus)*, or blood-vascular system *(Sclerostomum)*, and cause many serious diseases. One of the most dangerous of these to which man is liable, is Trichiniasis, caused by eating insufficiently-cooked pork, in the flesh of which the minute encapsuled Trichinæ are living (Fig. 151). The Acanthocephali are so-named, from a proboscis covered with hooks, by which they fix themselves in the intestines of their hosts—for the most part, lower Vertebrates.

11. The Plathelminthes also include some free forms, such as the marine **Nemertini**, unsegmented worms, which sometimes attain a length of thirty or forty feet (*Lineus*), and also the **Turbellaria**, for the most part very small creatures living in water or damp places. In both, the skin is covered with ciliated cells, which are absent in the other (parasitic) orders. The intestine is tubular and complete in the Nemerteans, but in the Turbellarians it is sometimes much branched, and always opens to the outside only by the mouth. The commonest forms are species of *Planaria* (Fig. 152 *b*) little flat, leech-like forms, which are to be found clinging to stones or creeping about in fresh-water ponds or streams. They are rarely longer than half an inch, but some marine forms attain a much larger size. On the other hand, there are fresh-water Turbellaria with a simple rod-like intestine (Fig. 152 *a*), which rarely exceed a line in length, and others, with no intestine at all, in which the food-particles are simply admitted by the mouth into the interior of the body, which is composed of soft cells.

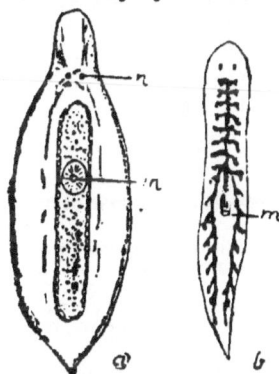

Fig. 152.—Turbellaria.

a, Mesostomum, with rod-like ; *b, Planaria,* with branched intestine ; *m,* the mouth ; *n,* nerve-ganglia with eye-spots.

12. The remaining orders of the Flat-worms are the **Trematodes,** and **Cestodes.** In the former, the intestine is a forked tube opening only by the mouth ; in the latter, it is entirely absent, and there is no cœlom in either. Organs of fixation in the form of suckers or hooks are always present. The Trematodes live either as ectoparasites on various animals, such as fishes and molluscs,—in which case they are well provided with hooks and suckers, and form only a few eggs, which they fix directly to their hosts, — or as entoparasites within various Vertebrates,—in which case they

generally have two suckers (genus *Distomum*), one round the mouth, the other on the ventral surface. The eggs are very numerous, for only a very small number of them can meet with the complicated conditions favourable to their complete life. They do not give rise at once to the Distome form, but, by internal budding, to intermediate broods, which differ from the adult in form, and are found in some different animal (Fig. 153).

Fig. 153.—Developmental cycle of *Distomum hepaticum.*—The liver-fluke of sheep. (After Leuckart).

a, the adult showing the position of the suckers, and the branched intestine; *b*, an egg with operculum, and contained ciliated embryo, with some unconsumed food-yolk.—This embryo gives rise to a "sporocyst" in which "rediæ," like *c*, are formed by internal budding. The redia gives rise, also by internal budding, to larval distomes—"cercariæ" *d*,—which loose the tail and after encystment attain the adult form, *a*.

13. Apart from the absence of the intestine, the Cestodes differ from the Trematodes in the formation of chains by budding, each segment in the chain or *proglottis*, resembling its neighbour, but differing from the original or head-segment, by the absence of organs of fixation. The segments have sometimes more, sometimes less capacity for independent life. The eggs formed within them do not at once develop into original head-segments, but into larval bladder-like forms (Cysticerci)—which are found in some different animal from the host of the adult—and these may, by budding, give rise to more than one head-segment. The adult chains are found in the intestines of all the classes of Vertebrates ; the cystic stages in the flesh, liver or brain of some animal, which serves as food for the host of the adult chains (Fig. 154). Thus the tape-worms of the carnivorous sharks pass through their cystic stages in the Teleosts,

on whichthey feed.   Through eating insufficiently-cooked beef, pork or
fish, man is liable to several forms of these parasites.

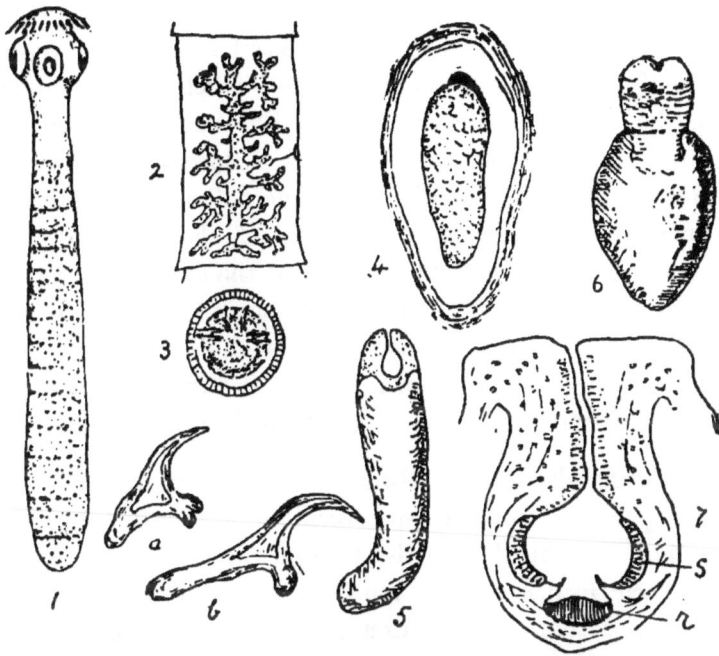

Fig. 154.—Developmental cycle of *Tænia serrata*. (A species of tape-worm
which occurs in the dog).   (After Leuckart).
    1, A young tape-worm composed of the head. with hooks (*a, b*), and suckers, and
a chain of immature segments ; 2, a mature segment, the branched uterus of which is
full of 6-hooked shelled embryos, 3 ; these gain access to the liver of the rabbit, loose
their hooks, become encysted, 4, and invaginated at one end which gives rise to the head
of the future tape-worm, 5 ; 6, a fully formed bladder-worm (*Cysticercus pisiformis*) ;
7, section of the head before its eversion ; *r*, the rostellum which carries the hooks ; *s*,
two of the suckers.

MOLLUSCA.

14. From the lowly-organised unsegmented worms, which we
have been considering, to a sub-kingdom like the Mollusca, which
contains some of the largest and most highly-organised of the
Invertebrates, seems a very long step, and yet it is to the Vermes,
and not to any of the other sub-kingdoms, that we must look
for any resemblance to the Molluscan structure.

The sub-kingdom derives its name from the softness of the tissues of the body, a peculiarity dependent on the fact that it is generally protected by a one or two-valved shell. Locomotion is effected by the specialised musculature of the ventral surface of the body—the so-called **foot**—which recalls, in many cases, the creeping surfaces of the Planarians, but is often curiously modified for other methods of locomotion.. The respiratory organs are generally gills, situated on one or both sides of the body and protected by a fold of the skin called the **pallium** or **mantle**; it is this portion of the skin which has the function of secreting the shell. One pair of excretory organs, similar in structure to the nephridia of the segmented worms, are present, but the nervous system is arranged on a plan entirely different from that of any of the Vermes studied.

15. Two subdivisions of Molluscs are recognized, which differ according to the amount of specialisation of the cephalic end of the body. The bivalve shell-fish, from the peculiar manner in which their food is secured and their almost sedentary habits, have none of that concentration of the sense-organs and of the nerve-centres into the "head," which we find in the other Molluscs. They are thus frequently called the **Acephala,** in contrast to the **Cephalophora.**

16. The Acephala form a single class—**Lamellibranchiata,** called so on account of the plate-like gills, which are present in all. It is chiefly a marine group, but representatives of both of the orders into which it is subdivided occur in fresh water, the most conspicuous being the fresh-water Mussels (*Unionidæ*), any one of which will serve as a type for the study of the class.

The shell, like the body, is symmetrical, the right and left valves being similar; it is only in attached forms of Lamellibranchs like the oyster, in which any great degree of asymmetry is to be observed. At the dorsal surface is to be noted the **hinge,** formed by an an uncalcified part of the shell, the **ligament,** ·

and often by teeth on the valves. In front of the hinge **are** the **umbones**, the first-formed parts of the valves; they generally incline forwards. Three layers may be seen in the shell, the outer brown **periostracum**, the thick prismatic layer formed by the activity of the thickened border of the mantle (Fig. 155), and the nacreous or pearly layer, secreted by the whole mantle surface. Pearls are the result of repeated layers of this substance being formed round foreign particles, which have got between the mantle and the shell. The mantle corresponds in form to the shell, hanging down right and left of the body: in many Lamellibranchs, the margins of its two lobes tend to unite, except in front, to allow egress to the foot, and behind at two points, the siphons, to allow water to get into and out of the mantle cavity, but this union does not occur in the Unios, and the siphons (Fig. 156), are mere specialised parts of the mantle-margin, which can be fitted together.

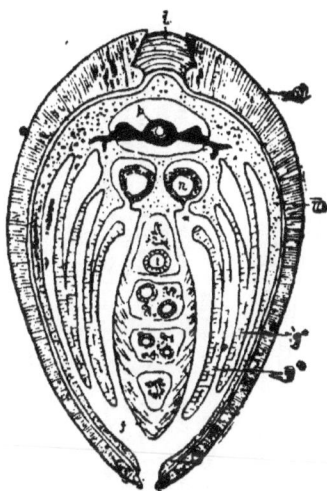

Fig. 155.—Diagrammatic transection of Anodon. (After Ludwig).

1, ligament; *sh*, shell; *m*, mantle; *m¹*, its thickened margin; *g¹*, *g²*, outer and inner gill-plates; *f*, in the mantle-cavity points to the foot, which contains portions of the intestine (*i*) and reproductive organs; *n*, is in the glandular part of the nephridium, * opening of its non-glandular part into the mantle-cavity.

Locomotion is effected by the ploughshare-shaped foot; muscular fibres are hardly developed elsewhere in the body-wall except along the border of the mantle, and especially at its anterior and posterior ends, where the **adductor** muscles which close the shell are situated.

Related to these muscular masses are the three pairs of nerve-ganglia—cerebral, pedal and parieto-splanchnic—with connecting nerve-cords. A pair of flat, triangular, tactile tentacles on # 230   HIGH SCHOOL ZOOLOGY.

each side of the mouth, a pair of otocysts in the foot, supplied by the cerebral ganglia, and a patch of olfactory neuro-
epithelium near the posterior adductor are the chief seats of
the special senses. In some forms, eyes are distributed along
the mantle-margin, but not in the Unios.

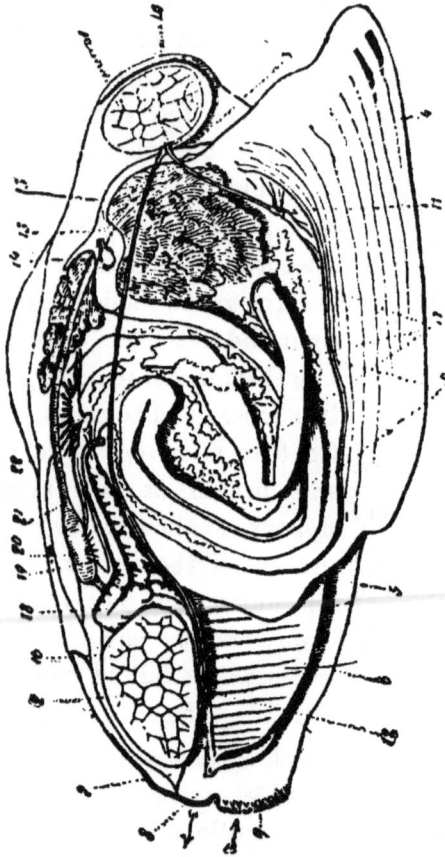

Fig. 156.—Dissection of a pond-mussel, *Anodon.*   (From Ludwig).
Gills of the right side and part of the body-wall removed:—1, anterior adductor muscle;
2, posterior adductor muscle; 3, cerebral ganglion; 4, foot; 5, mantle margin; 6, inner left
gill; 7, anus; 8, exhalent siphon; 9, inhalent siphon; 10, intestine; 11, pedal ganglion; 12,
parieto-splanchnic ganglion; 13, liver; 14, crystalline stylet; 15, stomach; 16, mantle-hood
with palp below; 17, reproductive glands; 18, nephridium; 19, auricle of heart; 20, ventricle of heart; 21, aorta; 22, pericardial gland.

The mouth lies under the anterior adductor and leads into
a stomach surrounded by a bulky liver, and possessing a cœcum
containing a "crystalline stylet" of unknown function. After
several turns within the foot, the intestine ascends, and in its
course to the upper surface of the posterior adductor, is envel-

oped by the heart. This organ is " systemic," driving the blood, which it has collected from the gills, forward and backward throughout the body. The most spacious part of the cœlom is the pericardium surrounding the heart; it communicates with the outside (the mantle-cavity) by means of a pair of nephridia (the Organ of Bojanus), which open into the pericardium anteriorly, and then turn upon themselves in such a way, that the distal non-glandular part of each tube lies above the proximal glandular part, and opens in nearly the same plane as the pericardial opening.

There are two gill-plates; each is formed of a number of vertical filaments attached to the side of the body-wall (as may be seen to the right of * in Fig. 155) and curved upon themselves, the inner series to the inside, the outer to the outside. The plate results from the union of filaments to their neighbours in front and behind; it is double, owing to the recurving of each filament, but the two layers are separated above, although the space between them is partly obliterated by junctions below.

Water is sucked through the inhalent or branchial siphon by means of ciliated cells on the gill-plates; the current sets through the surface of the plates into the spaces between the two layers, whence it is swept out through the exhalent or cloacal siphon, carrying with it the excreta from the kidneys and intestine. Solid particles contained in the water are swept forwards towards the mouth, and guided into it by ciliated cells on the tentacles.

The reproductive organs are situated in the foot, and the eggs undergo the greater part of their development in the interlamellar space of the outer gills, which are thus turned into brood-pouches. The distribution of the larvæ (Glochidia) is provided for by their escaping thence and fastening themselves in the skin of various small fish, where they undergo a resting stage before they attain their adult form.

17. The Unionidæ belong to an order Asphoniata distinguished by the absence of tubular siphons ; the oyster (*Ostrea*) and scallop (*Pecten*) also belong here, but they have only one (the posterior) adductor muscle, and no foot. Intermediate between these types is the sea-mussel (*Mytilus*), in which the anterior adductor and the foot are small. These forms are attached, not by the shell, but by horny "byssus" threads secreted by the foot.

18. A second order Siphoniata is formed for those in which the mantle cavity is closed except for the tubular s'phons behind, and an apertnre for the foot in front. Numerous minute fresh-water forms (*Cyclas*, *Pisidium*) belong here, but the bulk of the forms are marine. When the siphons are long, and can be retracted into the shell, there is a corresponding mark within the shell (F g. 157), as in the Sea-clams (*Mya* and *Venus*). Some of the forms burrow in sand, others in rocks (*Pholas*), or timber (*Teredo*).

19. The higher division of the Mollusca (the **Cephalophora**) presents a much greater range of form than do the Acephala; four Classes are recognized, each of them exhibiting important modifications of the typical Molluscan structure. Most of the species belong to one of these, the **Gastropoda**, called so on account of the development of the foot into a locomotive organ, generally a flat creeping surface, occupying the ventral aspect of the body.

Fig. 157.
Tellina grænlandica.

Very few Gastropods retain the bilateral symmetry which we see in the Acephala ; the most primitive forms—the Chitons— do, (a group, which of all the Mollusca, comes nearest to the Vermes), but in most a distinct asymmetry is present. This depends upon a separation of the vegetative from the animal organs, and the grouping of the former into a visceral mass or hump, above the head and foot, which are closely united (Fig. 158). The visceral mass is protected by a shell (similar in structure to that of the Lamellibranchs), which, in the simplest cases, is formed of a single conical piece ; in most Gastropods, however, the shell is spiral, and the visceral mass has, in the course of development, become twisted in the same direction, so as to cause the intestine *e. g.* to open anteriorly. This twist results in the suppression of the gill and nephridium of one side, and also in

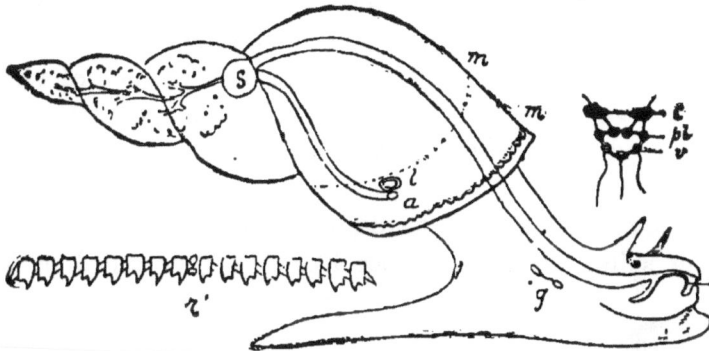

Fig. 158.—Diagram of Limnæa, to show the course of the intestine and the arrangement of the nerve-ganglia in the head.

r, floor of the mouth occupied by the radula; *r¹*, a row of lingual teeth; *g*, position of apertures of reproductive organs ; *m*, the free mantle margin ; *m¹* the line of fusion of the mantle with the body-wall bounding the lung in front ; *l*, the aperture of the lung; *a*, the anus ; *S*, the stomach, receiving tne tubes of the liver which occupies the apex of the shell : *C*, the cerebral, *pl*, the pleural, *v*, the visceral ganglion ; between the two pleural are the two pedal, and between the visceral, the single abdominal ganglion.

the asymmetrical situation of the heart. The direction of the twist determines which side of the body shall be affected ; it is sometimes towards the right (dexiotropous), but generally towards the left (leiotropous), in which case the organs of the right side are retained at the expense of those of the left.

The mantle-cavity, which is so roomy in the Acephala, is much restricted in the Gastropods, and is confined to the sides of the visceral mass. The mantle-margin is rarely free, but generally forms an enclosed space opening externally by an aperture or tube, which lies on the right side in left-twisted (leiotropous) forms.

Apart from the asymmetry referred to, the most striking dif-
16

ference is in the organization of the head. Not only are the chief nerve-centres and the sense-organs aggregated here, but there is developed a complicated mechanism in connection with the mouth, consisting of horny jaws and a lingual ribbon or *radula*, the surface of which is beset with teeth like a rasp or file, and which can be everted by special muscles.

20. Three-fifths of the Gastropods are adapted for breathing air, the mantle-cavity being altered into a lung and the gills being rudimentary (cf. VII, 12); they form an important order **Pulmonata**, a key to the structure of which is furnished by the pondsnail figured above. It belongs to a sub-order, the members of which *(Basommatophora)* have the eyes at the bases of the tentacles, and possess thin shells, which may be spiral like *Limnæa*, or spiral and dexiotropous like *Physa*, or coiled in one plane like *Planorbis*, or simply conical like *Ancylus* (Fig. 159). More numerous, however, are the land-snails and slugs which carry the eyes at the tips of the tentacles *(Stylommatophora)*, and which include shelled forms like *Helix, Zonites, Succinea* and forms with a rudimentary internal shell like *Limax*.

Fig. 159.—Shells of fresh water Gasteropoda.
Pulmonates,—1, *Helix*; 2, *Succinea*; 3, *Physa*; 4, *Ancylus*; 5, *Planorbis*. Prosobranchs,—6, *Paludina*, with the operculum in the aperture; 7, *Goniobasis*.

21. A second order of Gastropods—**Prosobranchiata**—includes, for the most part, marine forms, differing from the Pulmonata in possessing a gill in the mantle-cavity, and, usually, au operculum carried on the foot for closing the aperture of the shell. The ordinal name is derived from the fact that the respiratory organ is situated in front of the heart, as it is in the Pulmonata. Great variety of colour and form characterizes the shells of the Order, the Chitons, *e.g.*, having a shell formed of eight transverse pieces, the Limpets (*Patella*), a simple conical shell, while endless va-

rieties of spirals are to be met with in the Top-shells, (*Turbo* and *Trochus*), Olives, Cone-shells, Cowries (*Cyprœa*), &ç., &c. Some few Prosobranchiates, such as *Helicina, Valvata, Paludina*, are met with in fresh water (Figs. 159, 6 and 7).

Fig. 160.—Outline of a Heteropod (*Atlanta*). The foot is divided into three regions : pro-meso-, and meta-podium. On the mesopodium is a sucker, on the metapodium, an operculum.

Fig. 161. A naked Opisthobranchiate (*Doris*). A rosette of gills surrounds the anus.

22. In the **Heteropoda** we have a series of Gastropods adapted for a pelagic life, the foot being compressed into a fin, and the visceral mass and its protecting shell much reduced in size, so as not to interfere with the transparency of the creature (Fig. 160); and in the **Opisthobranchiata**, a series of marine forms, in which the shell is small or absent, but the gills generally project free from the surface of the body, and are situated behind the heart (Fig. 161).

23. The Gastropods are the only Class of Cephalophorous Molluscs which are represented inland ; the others are exclusively marine, and embrace comparatively few living forms. They are contained in three Classes, the **Scaphopoda** or Tooth-shells, with burrowing foot and numerous slender tentacles (Fig. 162 *b*), the **Pteropoda,** pelagic forms with or without a shell, but with the foot converted into two wing-like fins (Fig. 162 *c*), and the **Cephalopoda,** at present a small group in comparison with its development in past geological times (Fig. 162 *a*). To this class belong the most highly-organized Mollusca—the Cuttle-fishes (*Sepia*), Squids (*Loligo*), Octopus, &c.; in all of which the shell is reduced to an internal " cuttle-bone " or "pen." The foot is partly transformed into a circlet of ten or eight "arms" surrounding the mouth and carrying formidable suckers, and partly into a " funnel," which permits the water used in respiration to be forcibly ejected from the mantle-

cavity, and thus to be employed for swimming. Horny jaws
bound the aperture of the mouth ; and the nervous system and
sense-organs, which are protected by a cartilaginous endoskeleton, attain a degree of development not to be met with elsewhere among the Invertebrates. These forms have two gills

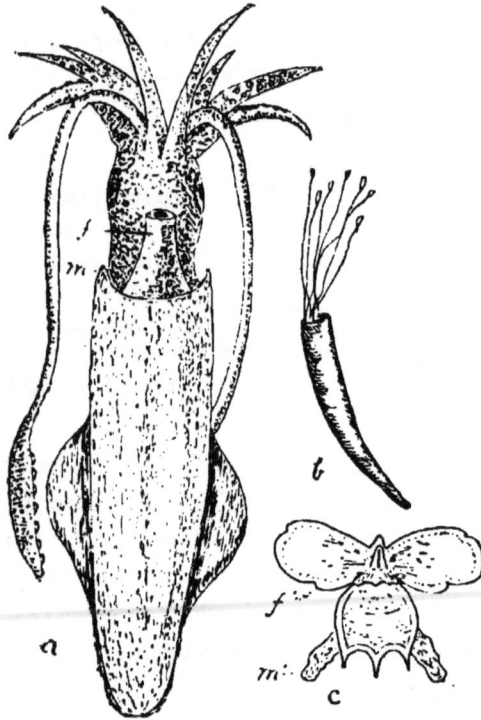

Fig. 162.—*a*, a Cephalopod (*Loligo*); *b*, a Scaphopod (*Dentalium*); *c*, a Pteropod
(*Hyalea*).
m, mantle-margin of Loligo ; *f*, the funnel ; *f*¹, the fins of Hyalea, between them the
mouth ; *m*¹, processes of the mantle-margin.

(Dibranchiata) in the mantle-cavity, but the Pearly Nautilus
(*N. pompilio*) has four (it is a Tetrabranchiate form), and it
further differs from the cuttle-fish in the number of the oral
subdivisions of the foot, in the presence of an external chambered
shell, and the absence of the ink-bag, (a very characteristic
organ of defence of the Dibranchiata, which furnishes the

"sepia" that they diffuse around them for concealment). Although fossil Dibranchiate forms are not uncommon (*Belemnites*), yet the bulk of the fossil Cephalopods belong to the Tetrabranchiata, the chambered shells of thousands of different species furnishing to the Palæontologist means of recognizing the relative age of the rocks in which they occur.

MOLLUSCOIDEA.

24. Two classes of animals require to be noticed in the present chapter, which have been associated together as the sub-kingdom **Molluscoidea**, partly on account of real, and partly on account of fancied affinities to each other and the Mollusca proper. These are the **Brachiopoda** and **Polyzoa**; the former, long considered to be related to the Lamellibranchs on account of the possession of a bivalve shell; the latter, on account of their forming colonies, formerly classed with the "zoophytes" to be described in the following chapter. There is no superficial resemblance between the two classes themselves, but zoologists have determined by studying the development of both, that they are not only related to each other, but also to the Vermes.

25. The Brachiopods are exclusively marine animals, comparatively few species of which survive to the present day; in past geological periods, however, they were extremely numerous, and have, therefore, much of the same interest attaching to them as the Trilobites and Tetrabranchiate Cephalopods. Most of the living species are found in the warmer seas; of the few that occur in the Gulf of St. Lawrence, *Rhynchonella psittacea* is perhaps the commonest. This species exhibits the characteristic inequality of the two valves of the shell (Fig. 163), the smaller (dorsal) valve fitting like a lid on the larger (ventral) valve, which also has a projecting beak permitting the passage of a short stalk by which the animal is attached. These valves

---

238   HIGH SCHOOL ZOOLOGY.

are secreted by mantle-lobes, which, of course, have likewise a

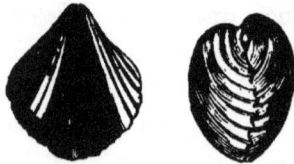

Fig. 163.—*Rhynchonella plena.* Chazy Formation.

totally different relation to the body, from what exists in the Lamellibranchs. Their edges are beset with setæ, like those of worms, and they contain processes from the cœlom in which the eggs are to be found. The animal is confined to the attached end of the shell (Fig. 164) the roomy mantle-cavity being occupied by a pair of coiled-up arms, fringed with tentacles, which take their origin on either side of the mouth. These arms (at one time supposed to be equivalent to the Molluscan foot—hence the name of the class—) are often supported by a

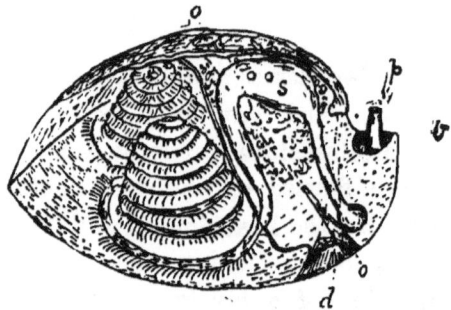

Fig. 164.—Diagram of Rhynchonella, as seen from the side with the shell partly removed. *a,* anterior; *b,* posterior or hinge area of shell; *p,* peduncle, *s,* stomach embedded in the liver and communicating with the œsophagus in front, and the intestine behind; d, opening, o, closing muscles.

calcareous endoskeleton; such is not the case in *Rhynchonella,* but in some of the fossil forms (Fig. 165), the complete skeleton is well preserved. It is obvious, then, that the arms are not for locomotion; their chief function is to bring food to the mouth, and this is effected by cilia on the tentacles, which create a current down the coils to the mouth.

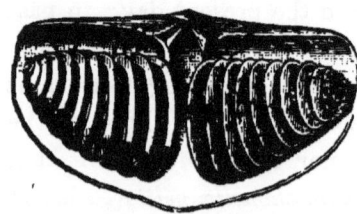

Fig. 165.—Internal surface of dorsal valve of a *Spirifer.*

One or two pairs of nephridia are present, more nearly resembling those of Vermes than of Mollusca. The intestine ends blind in Rhynchonella and other forms which have a hinge to the shell (**Testicardines**), but in

Lingula, (a form with a long flexible peduncle, which can displace laterally the upper valve of its shell owing to the absence of a hinge—Ecardines) the intestine turns forwards and opens near the mouth. Lingula is an example of a "persistent" type, as the generic characters do not appear. to have altered from Palæozoic times (Fig. 166).

Fig. 166.—*L. Quebecensis*: Levis formation.

Although the adults are attached forms, the larvæ are free; they are decidedly worm-like, being formed of three segments, the hindmost of which becomes transformed into the stalk, the foremost becomes much reduced, while the middle one gives rise to the body and mantle-lobes.

26. Like the Brachiopoda, the Polyzoa are bilateral animals, sedentary in their adult condition; they possess a circlet of tentacles about the mouth, and are protected by a shell secreted by the skin, but in other respects no resemblance is to be detected between them, except during developmental stages. The fact that the Polyzoa almost invariably form colonies by budding at once separates them from the forms heretofore studied; it is to this peculiarity that the class owes its name.

With the exception of *Cristatella* (Fig. 167), the colonies are permanently sedentary, being attached by an extensive or limited surface. According to the relative position of the buds they may be, in the former case, massive, encrusting or straggling, in the latter, foliaceous or arborescent (Fig. 168). The moss-like form of the colonies has gained the class the alternative name of *Bryozoa*. Each individual in the colony secretes a "cell" into which the head with the crown of

Fig, 167.—Colony of Cristatella, attached to stem of pond-weed.

tentacles may be withdrawn for protection, and these cells together with connecting tubes constitute the " cœnœcium." It may be of very different character in different forms, being sometimes gelatinous, but oftener horny or calcareous. The colonies may reach a considerable size, but the individuals rarely exceed one or two lines in length.

Most of the Polyzoa fall into two orders, which are nearly co-extensive with the Fresh-water and Marine forms respec-

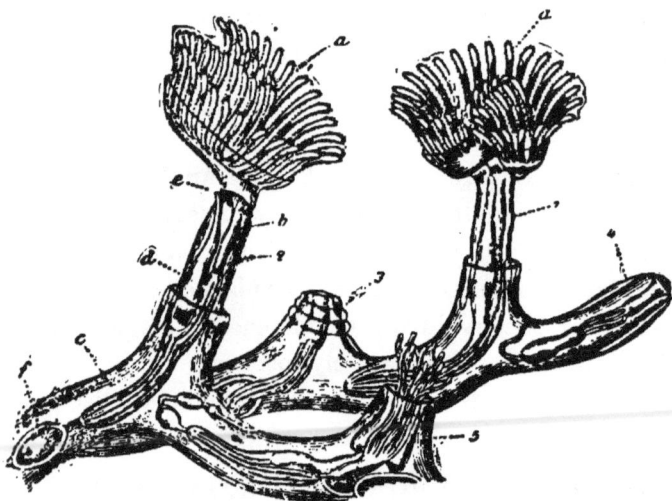

Fig. 168.—Portion of a colony of Plumatella enlarged.
1 and 2, expanded individuals ; 3, 4, 5, individuals in various stages of retraction into the cells. a, the lophophore ; b, œsophagus ; c, stomach ; d, intestine ; e, anus ; f, statoblast, attached to base of stomach.

tively. The former, which are abundant in our ponds and streams, have a horse-shoe shaped row of the ciliated tentacles and a cover (epistome), which closes the mouth (**Phylactolæmata**); the latter (**Gymnolæmata**)—of which *Paludicella* is the only Fresh-water example—a circlet of tentacles and no epistome.

Some idea of the form of the intestine and of the mechanism of retraction into the cell may be formed from Fig. 168. The nervous system is in the form of a ganglion between the apertures of the intestine, and there are no special sense-organs.

Neither are special circulatory or respiratory organs present, but a pair of excretory tubes put the cœlom in communication with the outside. In addition to increase by budding, the Fresh-water Polyzoa give rise to new colonies by winter-buds or **statoblasts,** which are protected by a double horny shell (often ornamented with hooks—*Pectinatella, Cristatella*—(Fig. 169). These float up to the surface in Spring and give rise at once to new colonies by budding.

Fig. 169.—Statoblast of *Cristatella.* × 25

In the marine forms, the colonies are frequently *polymorphic,* special individuals being modified for prehension alone, others, so as to act as brood-pouches for the developing eggs. Fossil remains of such as had calcareous cœncœcia (*e.g. Fenestella*) are abundant from the Silurian strata upwards.

CHAPTER IX.

THE REMAINING INVERTEBRATE SUB-KINGDOMS.

1. Four sub-kingdoms, the Echinodermata, Cœlenterata, Porifera, and Protozoa remain to be referred to; with the exception of the last, however, they have but few fresh-water representatives. In place of the bilateral symmetry of the foregoing sub-kingdoms, a radial symmetry is often more noticeable. On this account the groups in question were at one time known as the "Radiata," but it must be understood that bilateral symmetry may co-exist with the radial.

2. The Echinoderms, exclusively a marine group, receive their name from the general presence of an exoskeleton formed of more or less regular calcareous plates in the skin (Fig. 170), which carry protecting spines. As a rule the body is formed of five similar rays or "antimeres" grouped round a central axis. The intestine is usually complete and contained in a

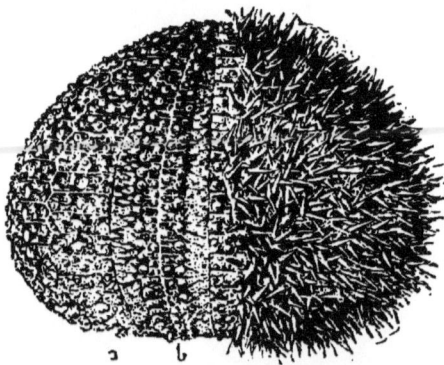

Fig. 170.—*Echinus esculentus.* ½.
Half of the shell is stripped of the spines, showing the double rows of imperforate plates, *a*; and of those perforated by the tube-feet—*b*.

spacious cœlom, from which there is separated off during development a system of blood-vessels, and also a characteristic system of water-vessels. The chief stems of the latter answer

Fig. 171.—*Pentacta frondosa.*
(U. S. F. C.)
A Holothuroid with tentacles expanded, and tube-feet protruded; the latter are arranged in distinct rows.

to the rays, and are provided with reservoirs, from which water can be forced into certain short processes of the stems, known as "tube-feet" (Fig. 171), which are thus caused to project from the surface of the body, so as to act as locomotive organs.

3. The larvæ are extremely unlike the adults, the developmental stages recalling in many respects those of a remarkable worm-like animal (*Balanoglossus*), the organization of which points to its being a very primitive form, presenting as it does, points of contact with several higher sub-kingdoms.

4. Of the various classes into which the Echinoderms are subdivided, the **Holothuroidea** are characterized by cylindrical form and a soft skin (Fig. 171). The tube-feet are often confined to a ventral surface so that the animals are then distinctly

Fig. 172. —Sand-dollar from above. — *Echinarachnius parma.* ½. The ten double-rows of plates of which the shell is constructed may be seen ; the upper ends of the perforated rows are modified into *petaloid ambulacra.*

Fig. 173.—*Pentaceros* from above.
A Bahaman Starfish.

Fig. 174.—Brittle-Star. *Ophiothrix fragilis.*

bilateral. The **Echinoidea,** on the other hand, may be globular (Fig. 170), or discoid (Fig. 172), but the skeleton is made up of regular rows of plates (generally twenty) some of which are perforated for the tube-feet. The Star-fishes **(Asteroidea)** and Brittle-Stars **(Ophiuroidea)** resemble each other in having a central disc and projecting arms, which bear tube-feet only on the ventral (lower) surface, but they differ in that the Starfish arms (Fig. 173) contain processes from the intestine, while this is not the case in the Brittle-stars (Fig. 174). Finally, the **Crinoidea** resemble the preceding in having a disc with arms, but the ventral surface with the mouth is uppermost, and the dorsal surface is temporarily or permanently fixed by a stalk (Fig. 175). Along with certain allied forms, which

Fig. 175.—Living Crinoid (*Pentacrinus*) with part of its stalk.

b. Upper surface of the calyx with the arms cut off, showing the mouth in the centre, and the furrows converging to it from the arms.

are now quite extinct, the Crinoids were much more abundant in the earlier geological periods, than they are at the present day.

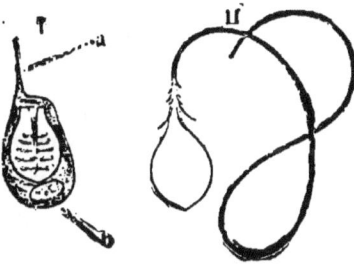

Fig. 176.—Thread-cells of Hydroid. I, undergoing development; II, with the thread protruded.

5. The Cœlenterata also are "radiate," but the parts are usually disposed in twos, or fours, or sixes; instead of the complete intestine and complicated cœlom of the preceding group, there is only one cavity which discharges the functions of both; peculiar modified cells (thread-cells or nettling organs, Fig. 176), take on a defensive function, by pouring an acrid poison into wounds inflicted by their microscopic barbs. From the tendency to form colonies (often plant-like in form) this group, which is almost exclusively marine, used to be called the **Zoophytes**.

Fig. 177.—*Hydra viridis*, attached to a weed, with buds in two stages of development. The expanded tentacles appear granular owing to accumulations of thread-cells. *a, a* thread-cell.

Apart from some singular pelagic forms which have comb-like ridges of cilia on the surface (**Ctenophora**), the Cœlenterates fall into two classes—the **Hydrozoa** and the **Actinozoa**. The former name is derived from the fresh-water Polyp, *Hydra* (Fig. 177), a cosmopolitan form very much simpler in its organization than its marine allies. It does not form colonies, but buds off individuals which soon become detached from the parent. Eggs are formed, during a short period of the year, in the wall of the two-layered tube which constitutes the body.

One end of the tube is closed, and serves for attachment, the other opens by the mouth, which is surrounded by tubular tentacles, capable of extraordinary elongation, and employed for seizing the minute animals on which they feed. Both layers of the body-wall (ectoderm and endoderm) take part in the f rmation of the tentacles. Hydra owes its generic name to its extraordinary power of recovery after injury, any fragment of an individual being capable of reproducing the rest. The marine Hydroids form colonies, often arborescent (Fig. 178), the connecting stems of which are always, and the individuals frequently, protected by a horny exoskeleton—the perisarc (Fig. 179). Eggs are formed in peculiarly-shaped individuals,

Fig. 179.—Diagram of Hydroid colony (after Allman) b, root or hydrorhiza: a, cœnosarc; a^1, perisarc; h, modified individual with reproductive buds (i) about to assume the form of locomotive medusæ like k; c, nutritive individual; e, mouth; g, tentacles.

Fig. 178—Hydroid colony. (*Obelaria gelatinosa*).

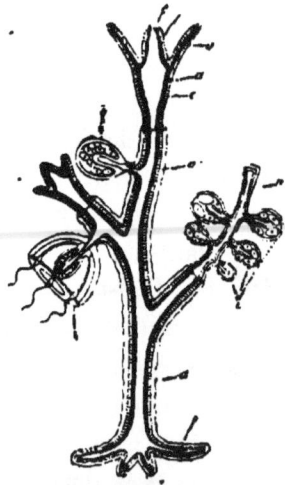

which, in some cases, differ very much from the ordinary polyp
in shape, and may even be detached and swim off as *Medusæ*
(Fig. 170). There is then an alternation between the Hydra-
form, multiplying by buds, and the Medusa-form multiplying by
eggs. Some Medusæ, however, the larger jelly-fishes (Fig. 180),
do not come from Hydroid colonies, but their eggs give rise to
tube-like larvæ, which undergo multiplication by division into
young Medusæ, related to each other like a pile of saucers. In
such higher forms of Hydroids an abundant layer of gelatinous
connective tissue (mesoderm) separates the endoderm from the
ectoderm.

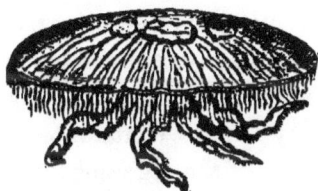

6. The class Actinozoa (exclusively marine) derives its name from
the Sea-anemone—*Actinia*—(Fig.
1º1). These creatures attain a
considerable size, and are often very
brilliantly coloured. They differ
from the Hydroids by having the
mouth-end of the tubular body

Fig. 180.—Higher Jelly-fish.
(*Aurelia aurita*).

turned in as a stomach-sac, which is connected by mesenteries or
septa to the outer tube (Fig. 182). The chambers between
the septa open above into the tentacles,
which are separated from the mouth by
an intervening disc. The Sea-anemonies
have no skeleton, and do not form
colonies, but allied forms give rise by
budding, or division, to colonies, which
may be arborescent or massive, and in
which a skeleton or **corallum** is a
marked feature. The corallum may
be confined to the axis of the common
flesh (cœnosarc), which unites the individual polyps, as in the Fan-corals and
the Red Coral of commerce (Fig. 183),
or it may invade the polyps them-

Fig. 181.—A Sea-anemone (*Actinia*). a, the mouth; b, the disc;
c, the tentacles; d, margin of the
disc; e, the wall; f, the base.

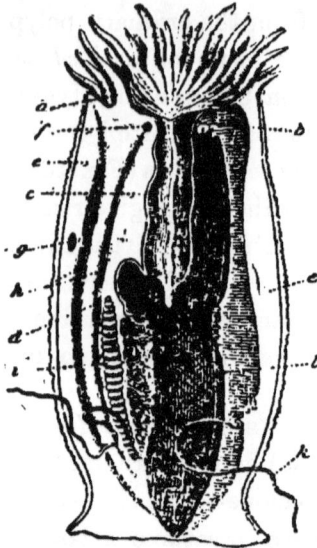

Fig. 182.—Diagram of Actinia.
(From Ludwig).

a, tentacle ; b, mouth ; c, stomach-sac ; d, its opening into cœlenteron l ; c, septum ; e¹, secondary septum ; f, g, apertures in septum ; h, muscular slips ; i, reproductive organs k, mesenterial filaments.

selves. In the latter case, it may be only in the form of detached calcareous spicules, but oftener the spicules unite into a continuous structure, which penetrates the wall alone (*Tubipora*, Fig. 184) or even the septa of the polyps, forming thus a " cup-coral " (Fig. 185). Abundant fossil examples of such cup-corals occur in the Palæozoic strata of Ontario, but the living representatives of the group are chiefly confined to the warmer seas of the present day, where many species contribute to the formation of the different kinds of coral-reefs.

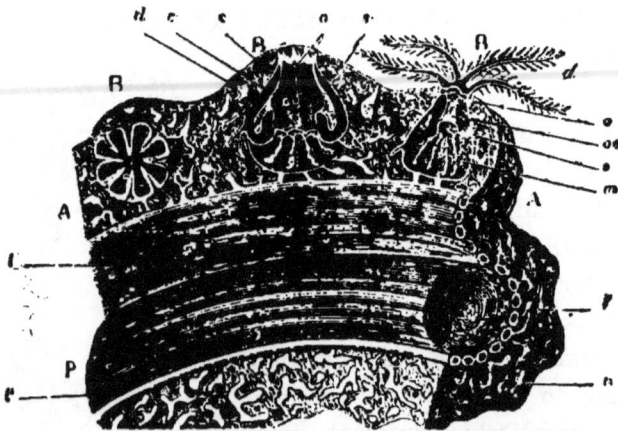

Fig. 183.—Red coral. (*Corallium rubrum*).

P, the calcareous axis or sclerobase. A, cœnosarc investing the axis and containing the individual polyps, B, these are in different stages of retraction ; d, tentacles ; œ, stomach-sac ; m, mesenteries. The cœnosarc is cut through and turned back, so as to show the canals (l and n) which traverse it, and the axis which it covers.

7. By many zoologists the **Porifera** or Sponges are regarded as Cœlenterates, chiefly distinguished by the absence of thread-cells; they have so many other peculiarities, however, that it is convenient to consider them apart. Although one family (*Spongillidæ*) is confined to fresh-water, and is abundantly represented in our lakes and streams (Fig. 186), yet it does not

Fig. 184.—Organpipe coral.
(*Tubipora musica*).
Some of the polyps are expanded.

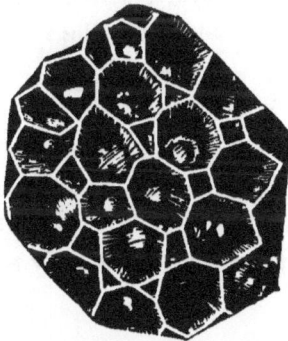

Fig. 185.—*Michelinea convexa.*
Devonian.

Fig. 186.—*Spongilla lacustris.*

contain conspicuous forms, and the word "sponge' is most frequently connected with the marine animals, whose skeletons we use in every day life. Yet those are derived from but one

17

(*Spongidæ*) of very many marine families, and in fact from comparatively few species of it. The skeleton in other families is
either formed of calcareous spicules (**Calcarea**), or of siliceous and
horny material, together or separate (**Non-calcarea**); it is only
in one family of the latter that there is no skeleton.

8. The name Porifera is derived from the presence of
numerous smaller and larger apertures on the surface of the
living sponge, respectively called "pores" and "oscula;" water,
carrying particles of food, is caused to stream through a more
or less .complicated system of cœlenteric canals, of which the
pores are the inhalent, and the oscula the exhalent apertures.
Certain parts of the canals—the ciliated chambers—are lined
by tall ciliated cells of characteristic shape—"collar-cells" (§ 16)
—and it is especially those which are active in bringing about
the current. They are equivalent to the entoderm of the
Cœlenterates, the rest of the canal-system and the outer surface
being clad with flattened ectodermal cells, while the soft, gelatinous, connective-tissue between these layers, which constitutes
the bulk of the "flesh" of the sponge, belongs to the mesoderm.

In some cases the cœlenteron, or gastrovascular cavity, has
only a single "osculum" (Fig. 187), and then the sponge may
be compared to a single Cœlenterate polyp, but generally it is a
more complicated canal-system with numerous oscula, and the
resemblance to a Cœlenterate colony is lessened by the absence
of symmetry and of constancy in the form of the body and the
position of the oscula.

9. Sponges are only possessed of locomotive powers in their
early larval stages; they afterwards attach themselves, and
possess, therefore, but little muscular tissue. Special sensitive
cells are present in the skin, but there are no sense-organs, such
as are found in the higher Cœlenterates. In addition to the
formation of new individuals from eggs, a separation of buds,

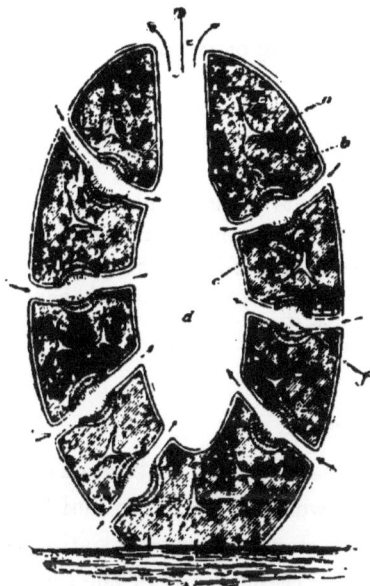

Fig. 187.--Diagram of a Calcareous sponge, with a single osculum.

a, ectoderm; *b*, mesoderm, with triradiate spicules; *c*, lining of gastro-vascular cavity; *f*, ciliated chambers lined with "collar-cells;" *e*, osculum. The arrows indicate the direction of the current inwards through the pores, and outwards through the osculum.

or even of parts of the living sponge, may also give rise to them. A peculiar kind of budding occurs in the fresh-water Sponges, which recalls the formation of the winter-buds of the Polyzoa, as it takes place under similar conditions. The buds are called statoblasts or " gemmules " and are protected by characteristic spicules (Fig. 188).

10. Fossil remains of sponges are abundant in the earlier formations, but they reached the height of their development during the upper Secondary period.

11. The Calcarea are chiefly minute forms found in shallow water; the same is true of the fleshy Non-calcarea—*Halisarca*,—but most of the

others attain a considerable size. Some of those with purely siliceous skeletons, like *Euplectella* and *Hyalonema*, are most beautiful objects; they occur in the depths of the ocean, anchoring themselves in soft mud by a wisp of glassy threads. In these forms, there are triaxial spicules in addition to the fibres, but in many sponges from shallow water, the spicules alone constitute the skeleton, which may

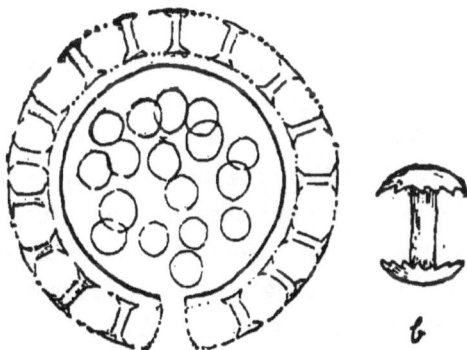

Fig. 188.—Diagram of gemmule of a fresh-water sponge, showing the coating of amphidisks and the aperture; *b*, an amphidisk of *Ephydatia*— the river sponge.

then be cork-like (*Suberites*) or friable in consistence (*Spongilla*). One genus—*Cliona*—has the singular habit of boring by means of its spicula into limestone and shells. The sponges of commerce, which come chiefly from the Mediterranean and the Bahamas, have the skeleton of **spongin,** either entirely free from foreign matter (in the best Turkey sponges— *Euspongia officinalis*), or else somewhat coarser in texture, from the building of siliceous spicules and fragments of sand into the horny fibre (horse-sponges—*Hippospongia*). Although much variety of form is to be met with in this family, still wider ranges in this respect are to be found in other families—not only massive, but tubular, funnel-shaped, dendritic and encrusting forms being met with.

THE PROTOZOA.

12. All the foregoing sub-kingdoms have one feature in common, which is not shared by the lowest forms of animal life; they pass through certain stages of development consisting of the segmentation of the egg, and the arrangement of the resulting spheres into a blastoderm, which always possesses at least the two primary layers, ectoderm and entoderm. The tissues, also, in all are the result of the further division and differentiation of these spheres; but in the sub-kingdom Protozoa, to which we now proceed, development does not take place in this way, and even the most highly organized forms are **unicellular,** the organs of the body being differentiated out of parts of one and the same cell. These important differences are expressed by uniting the foregoing sub-kingdoms under the one designation **Metazoa.**

13. The Protozoa are generally microscopic in size, and, with the exception of some parasitic forms, are confined to the sea and to fresh-water. Four classes are distinguished—the **Sarcodina, Sporozoa, Mastigophora,** and **Infusoria.** The two latter are often spoken of as Flagellate and Ciliate Infusoria, on account of their characteristic locomotive organs; neither flagella nor cilia are present in the first two classes; indeed the Sporozoa, being parasitic forms are destitute of any locomotive organs, and, in the

Sarcodina these are of the nature of "pseudopodia," *i.e.*, less permanent processes of the sarcode or cell-plasma than the flagella or cilia, and either thread-like or lobate in form.

14. In passing from the study of the lower to the higher Protozoa we shall have an opportunity of seeing to what extent "organization" can be carried within the limits of a single cell. The Sarcodina offer comparatively little of such differentiation; food-particles absorbed by the pseudopodia are conveyed into the softer cell-plasma at any point of the surface, and the undigested remains are similarly thrust out anywhere. Reproduction is effected chiefly by division into two or many cells, in some, a resting stage of "encystment," preceding division.

15. Among the simplest is the genus *Amœba* (Fig. 189), called so on account of its ceaseless change of form, the endoplasm, which contains the nucleus, is more diffluent than the ectoplasm which forms the pseudopodia and contains the "contractile vacuole," a structure which appears to discharge the function of an excretory organ in these and other Protozoa. The Sarcodina are not all naked like the Amœba; in the order to which it belongs (**Rhizopoda,**—called so on account of the root-like branches of the pseudopodia in many), most of the forms secrete shells (**Thalamophora**), which may be perforated all over for the escape of thread-like pseudopodia, or have merely one aperture through which similar processes escape or lobate ones like the Amœba's. The latter is the case in *Arcella* and *Difflugia* (Fig. 189, 2 and 3), while in *Euglypha* and others the processes are thread-like (Fig. 189, 4). These are fresh-water forms, in which the shells are formed of chitinous matter, or of foreign particles cemented together, but the marine forms—generally called **Foraminifera**—have for the most part a calcareous shell. This may be imperforate or perforate, and is generally composed of numerous chambers disposed in variously formed spirals (Fig. 190). Although of small size, these **Foramini**

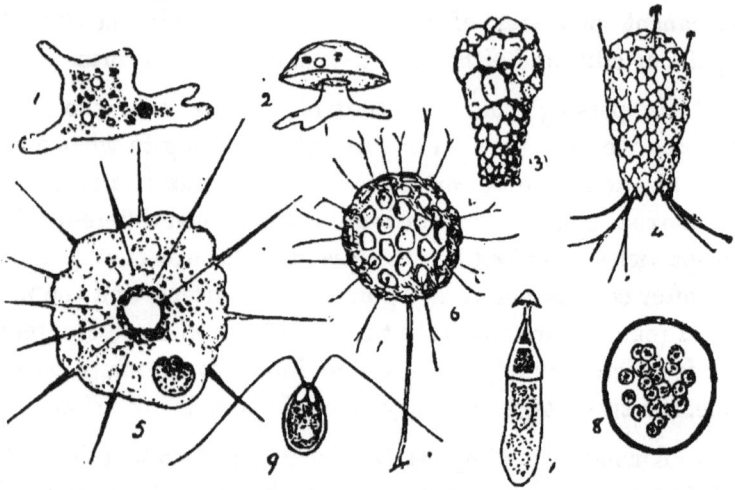

Fig. 189.—1, Amœba ; 2, Arcella; 3, Shell of Difflugia ; 4, Euglypha ; 5, Actino-
phrys, a naked Heliozoon with axial filaments in the pseudopodia ; 6, Clathrulina, a
stalked shelled Heliozoon ; 7, Gregarinid, from an insect-larva ; 8, Encysted Gregarina,
the contents transformēd into spores ; 9, Spore from a Myxosporidium.

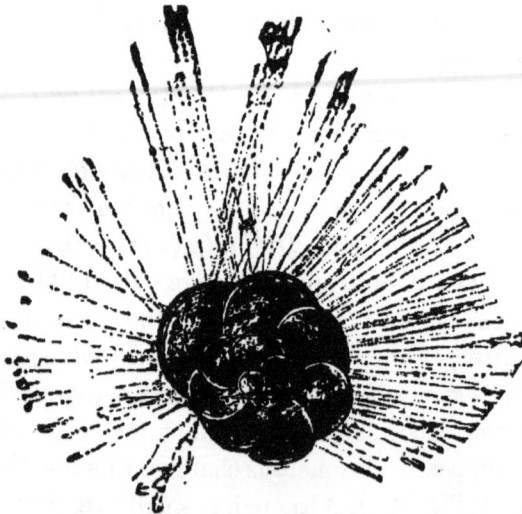

Fig. 190.—Living Foraminifer : *Rotalia veneta.*

fera have a very great geological importance, rocks of the chalk and other formations being often formed largely out of the remains of their shells. A deposit of 'a similar nature is being formed at the present day at the bottom of the Atlantic, dead shells of *Globigerina* and similar pelagic forms being rained down upon the bottom.

16. The pseudopodia in the Rhizopoda may flow together round a particle of food as represented in Fig. 190, but in the remaining orders of Sarcodina, they are less subject to alteration in form, rarely coalesce, stand out radially from the body, and are sometimes strengthened by an axial filament. These orders are the **Heliozoa** and the **Radiolaria**, the former a small group, chiefly of fresh-water forms, sometimes naked, sometimes with a spicular or perforated shell (Fig. 189, 6); the latter, a marine group containing numerous forms with siliceous skeletons, which offer the most surprising beauty and wealth of form (Fig. 191). The Radiolaria are also important from a geological, point of view, for deposits of "infusorial earth" are found consisting almost entirely of their shells, and similar deposits are being formed in some parts of the ocean at the present day. The body is more differentiated than in the Heliozoa, the endo-

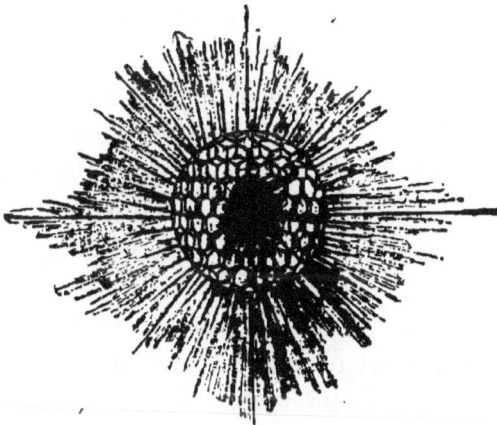

Fig. 191.—Living Radiolarian :—*Heliosphœra actinota.*

plasm being separated off by a special "central capsule" from
the ectoplasm. In the latter there are often found minute yellow Algæ, which appear to live "symbiotically" with the Radiolaria.

17. The **Sporozoa** are distinguished from the Rhizopoda not
only by the absence of pseudopodia, but by the presence of a well-
marked cuticle which limits the contractions of the protoplasm ;
they reproduce by spores, formed by the simultaneous division
of the plasma of an encysted individual into a multitude of globular bodies, which eventually acquire characteristically-shaped
shells (Fig. 189, 7,8). They are all parasitic, some of them being
intestinal parasites of the Invertebrates (*Gregarinidæ*), others,
so-called cell-parasites, which multiply in epithelial or blood-cells
of both lower and higher animals, and others, finally, ectoparasites of fish, being found on the gill-filaments. The spores of
the last forms have singular lasso-like organs of attachment
(Fig. 189, 9.)

18. The name **Infusoria** ("occurring in infusions"—an indication of the saprophytic life of many of the forms—) often
employed for the two remaining classes, was at one time used
to cover a host of microscopic creatures, to many of which, such
as the Rotifers, we have already given some attention ; it is
now restricted to the higher Protozoa, and frequently to those
which are ciliated, the class name **Mastigophora** being reserved
for those which have flagella. As they undoubtedly include
the simplest forms, which touch both upon the Sarcodina and
the Vegetable Kingdom, they may be mentioned first.

19. Many of the most minute animals belong to the flagellate
Infusoria (Fig. 192), various monads—*e.g.*, *Monas termo*—hardly
reaching 1-2000th of an inch in length. Such a form presents
definite specific characters, however, and retains a definite shape.
There can hardly be said to be a mouth ; rather there is a
vacuole at the base of the flagellum, which takes up food-

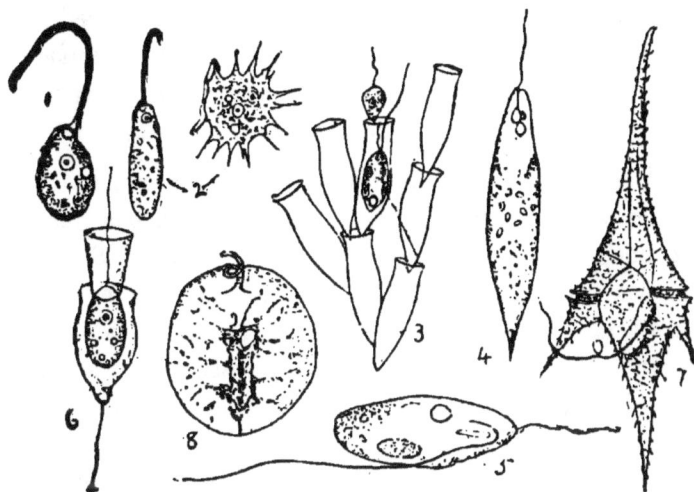

Fig. 192.—Types of Flagellate Infusoria.

1, Monas termo; 2 Ciliophrys in its two stages; 3, Dinobryon, one of the shells contains an animal which has given rise by budding to a new individual; 4, Euglena; 5, Anisonema; 6, Salpingœca with delicate "collar" standing up round the flagellum; 7, Ceratium; 8, Noctiluca.

particles whipped into it. A simpler method of obtaining food occurs in *Ciliophrys*, which throws out pseudopodia for this purpose, and passes into a flagellate stage for locomotion. Many of the monads form colonies; *Dinobryon e.g.*, not only does so, but possesses another peculiarity which is found in many —the ability to secrete a shell. The most interesting forms with regard to the process of nutrition are *Euglena* and its allies. They are active locomotive forms provided with a mouth, but the latter serves merely to get rid of the contents of the con-tractile vacuole, fluid nourishment being taken up through the cuticular body-wall, as in the lower plants. They further re-semble plants, by possessing coloured bodies which have the same physiological significance as the endochrome of the Algæ, *i.e.*, they serve in the presence of light to decompose carbonic acid, so that the animal secures part of its carbonaceous food in this way. Various forms like *Volvox*, usually regarded as plants,

might be equally well placed along with these coloured Flagel-'
lata. There may be more than a single flagellum ; *Anisonema,*
e.g., has two, one of which is stout and used for springing ;
others have four or six alike in character.

20. A remarkable group, the, members of which resemble
the collar-cells of the sponges in form (§ 8), has representatives both in fresh and in salt water. *Salpingœca,*
which has both the collar and a case, may serve as an example of it. Finally, reference must be made to two forms
of marine Flagellata, which are interesting on account of
assisting in producing the phosphorescence of the sea, in which
they occur in great numbers. *Ceratium* may be taken as a
type of the larger group, marked by the possession of a resist-
ent case, and by a second flagellum situated in a groove, which
looks like a row of cilia when in movement. *Noctiluca,* on the
other hand, attains a much larger size ; it is peach-shaped, with a
tentacle at the head of the groove which lodges the mouth, and
a flagellum and tooth within the groove. It is distinguished by
the reticular character of the protoplasm within the cuticle.

21. As a type of the Infusoria proper or **Infusoria Ciliata**, one
of the largest and commonest forms, *Paramœcium*—the slipper-
animalcule—found everywhere in water containing decaying organic matter, may be studied. It attains the size of 1-100th of
an inch, and is, therefore, visible to the naked eye. Mouth and
anus are both present, as they are in all, with the exception of
s me parasitic forms (*Opalina*), and there is a distinct œsophagus
leading inwards from the mouth to the endoplasm. The body is
uniformly covered with similar cilia (it belongs to the order **Holo-
tricha**) and there are, in addition, certain thread-like structures—
" trichocysts "—peculiar to this family, which can be thrust out
from the cuticle and seem to have a function similar to that

of the nettling-organs of the Cœlenterates. Two functions are discharged by the cilia; they bring food towards the mouth, and they serve for locomotion, but the contractile ectoplasm assists in the latter function. Two contractile vacuoles are present in this genus, which discharge their contents by radial tubes. Within the diffluent endoplasm may be seen food-particles circulating, which are being subjected to its digestive action; the nucleus is also situated there. This organ is peculiar in the

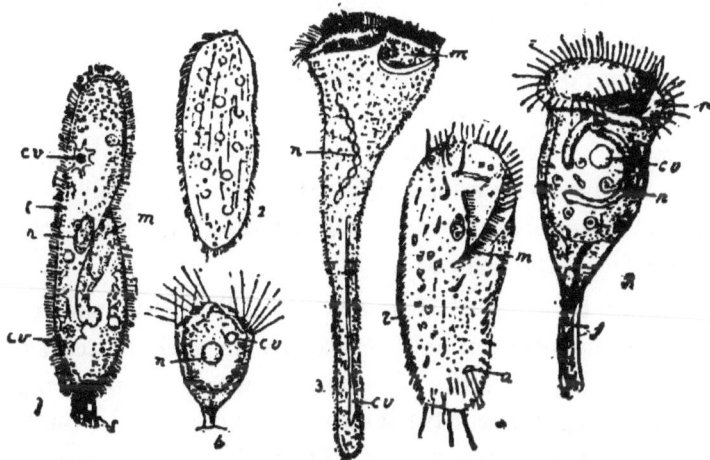

Fig. 193.—Types of Ciliated and Suctorial Infusoria.

1, Paramœcium; 2, Opalina; 3, Stentor; 4, Onychodromus; 5, Vorticella; 6, Acineta; *cv.* contractile vacuole; *m*, mouth; *v*, vestibule, leading towards mouth; *n*, nucleus; *t*, trichocysts; *a*, anus; *f*, contractile fibre in stalk; *r*, cilia of "right" border.

Ciliata and presents many varieties of form (Fig. 193); it consists of two parts, a larger and a smaller (the nucleus and the micronucleus); both of these play an important part in the method of multiplication by fission, which is so common in this group. They are also active in conjugation, a method of reproduction which occurs in other groups of Protozoa as well as in the Algæ.

22. The trumpet-animalcule (*Stentor*) may be taken as an example of another order (Heterotricha) where the cilia which

surround the mouth—" adoral "—are different from those clothing the rest of the body ; there is, as it were, a division of labour
between the locomotor and the nutritive cilia.   A third order
(**Hypotricha**) has the cilia confined to one surface, which therefore becomes a locomotor surface, and some of the cilia on that
are generally converted into hooks or spines as in *Onychodromus*,
while the fourth—**Peritricha**—which contains numerous species
tending to be attached and to form colonies, have the cilia confined to the neighbourhood of the mouth.   The Peritricha are,
however, by no means motionless ; they generally have a contractile fibre in the stalk of such forms as the Bell-animalcule
(*Vorticella*), but they may be free or parasitic forms swimming,
or creeping by their adoral cilia, or they may simply be able
to retract themselves into a shell or case, in virtue of the
contractility of the cell-body.

23. Many of the Peritricha are found living on the surface
of various aquatic animals, especially in such places as the
gills, where they have the advantage of a continuous change of
the surrounding water.   Such also is the case in the **Suctoria**,
a group of Infusoria which begin life with cilia, but afterwards
settle down and replace them by suckers or tentacles through
which, in the absence of a mouth, they take in their nourishment.   The suckers are tubular, and communicate with the
endoplasm, to which they carry the juices of the other Infusoria
on which they prey, or of the aquatic animals to which they
are attached.   Both fresh-water and marine forms belong to
this group, as well as to the other orders of Ciliata, but the
fresh-water forms are naturally best known.

CHAPTER X.

GENERAL PRINCIPLES.

1. In the preceding Chapters the principal forms of Animal Life have been reviewed, a particular standpoint having been selected in each of the chief subdivisions of the Animal Kingdom. An accessible and, where possible, a primitive type has been somewhat carefully examined, and the modifications of form then traced throughout the sub-kingdom or class to which it belongs. The examination has been confined to the more obvious characters of the adult organism, but occasional references have been made to minute structure, and to the developmental stages through which the adult form is reached. It has been made apparent that zoologists classify animals according to the degree of resemblance which they exhibit in these respects, and that, therefore, classification is a Synopsis of the results of such structural studies.

2. While it has been chiefly with this—the MORPHOLOGICAL— aspect of Zoology that we have hitherto been dealing, other topics have been incidentally touched upon. It has been made evident that there is the closest relation between the form of organs, and the uses to which they are put (whatever may be the explanation of this relation), that the various kinds of animals are limited to certain parts of the earth's surface, that the forms of life at present on the earth are different from those which occupied it in past times, and that there is the closest connection between animals and their surroundings. The present chapter will be reserved for a more systematic discussion of such topics as these, and an endeavour will be made to

state briefly the general principles which have been arrived at
by zoologists in regard to them, and to indicate the relations to
each other of the various aspects of zoological study.

3. Zoology is one of the two divisions of BIOLOGY (that
science which has for its subject-matter all things which have
or have had " life," in contrast to the lifeless objects of inorganic nature), it is, in fact, the Biology of animals, while Botany
is the Biology of plants. It may be asked why, in the face of
the palpable differences which exist between plants and animals, it should be necessary to have a common term for the
study of the phenomena manifested by both. The answer is,
that the phenomena manifested by living matter in both kingdoms present such a striking contrast to those manifested by
not-living matter, that the study of the difference between them
becomes a question of the first importance. It is not meant
that the matter which enters into the composition of an organism is different from that of which lifeless things are made, for
there is a constant exchange of matter between the living and
the lifeless world. An. organism has been well compared to a
wave formed by an obstruction in a rapid, the shape of which
is approximately constant, but the particles composing which
are incessantly changing. So the living organism is constantly
taking into it matter from the inorganic world. around it, and,
as constantly, parting with matter to its surroundings (as we
see in its relations, for example, to the gases of the surrounding atmosphere). Nor is it meant that matter in an organism
conducts itself in any exceptional way, for wherever it has been
possible to follow the transformations of matter and of energy
in the living plant or animal, it has been found that these take
place in harmony with laws which have been deduced from the
conduct of matter in the inorganic world. What then are these
manifestations of "life" and the properties common to plants
and animals, which induce us to regard Zoology and Botany as
parts of a more comprehensive science—Biology?

4. They may be stated as follows :—(1) Life is always associated with protoplasm, which has accordingly been termed the "physical basis of life." Even in its simplest forms, this substance is undoubtedly extremely complex from a chemical standpoint, while in its more differentiated forms, it contains temporarily within it innumerable "organic compounds" which form a considerable part of the subject matter of Organic Chemistry. All forms of protoplasm are constituted largely of "proteids" —complex compounds of Carbon, Hydrogen, Oxygen and Nitrogen, with small proportions of Phosphorus and Sulphur, but they also form within them Carbon compounds of greater simplicity, such as starch, sugar, fat, etc., which, however, do not occur in nature except as the products of the organic world. (2) The life of protoplasm (as manifested, for example, by irritability and contractility) is accompanied by its partial destruction, and this involves repair by the incorporation and assimilation of new matter, or, in other words, the ingestion of food and changes whereby this food can be rendered available for replacement of the material destroyed. (3) If the income of food is in excess of the expenditure by waste, growth results ; but each individual organism has a limit of size which it cannot exceed. A process succeeding the attainment of the full size in the simplest organisms is that of division, whereupon the life of the individual terminates and a new generation composed of two new individuals replaces it. Similar phenomena of death and reproduction occur also in the higher organisms. Nothing like these characters is to be met with in the Inorganic world, but this is not the only reason for studying the life of plants and animals together, for, apart from the circumstance that there are certain organisms (IX, 19) which can hardly be said to have struck out on either of the diverging paths which lead to plant and animal peculiarities, there is so much interdependence between the two, that a knowledge of the life of either is not complete without considering both, and indeed this interdepen-

dence is in part due to the differences which we shall refer to hereafter. However great these differences are, it must, nevertheless, be understood that the resemblances between plants and animals are such, that the methods of Zoology are the same as those of Botany, and the aspects of the study of the Biology of plants, the same as those of the Biology of animals.

VARIOUS ASPECTS OF BIOLOGICAL STUDY.

(1) MORPHOLOGICAL.

5. This division of Biological study involves, as we have seen, questions of form and structure. It admits of sub-division according as these questions deal with the adult, or with stages in the development of the adult : **Anatomy**, in its widest sense, being the term reserved for the former group of questions, **Embryology** or **Ontogeny** for the latter. In a narrower sense, Anatomy is restricted to such structural points as can be studied without the aid of the microscope, while **Histology** is employed for the study of the finer details of the tissues, and **Cytology** for those of their component cells. It is evident that questions, other than those of pure form, must be inseparable from the studies which have just been styled morphological. The changes of form, for example, in individual cells are merely the expression of physical and chemical changes taking place within them, so that Cytology might as legitimately be referred to the following aspect of Biological study.

(2) PHYSIOLOGICAL.

6. Morphology is sometimes described as statical Biology, because it involves the notion of rest; **Physiology**, on the other hand, as dynamical, because it studies the work-

ing of the plant or animal as a machine, its object being to follow the transformations of matter and of energy within the living organism. Although it is a comparatively easy task to trace how the energy locked up in fuel is made available by the steam-engine, it is an infinitely more difficult one to trace how the energy locked up in food becomes available as muscular work; yet there are other activities of the body which lend themselves far less easily to measurement than does that of the muscular system.

It is, then, the various activities of the organism which Physiology discusses. **Function** is the term employed for the rôle or duty discharged by any part of a plant or animal, and the part discharging a particular duty is termed an **organ** these terms are thus used correlatively. We have seen that different kinds of activity such as nutrition, locomotion and reproduction may be performed by apparently undifferentiated or little-differentiated protoplasm; it is therefore necessary to beware of supposing that **organization** or differentiation into separate organs is necessary for the display of the activities of life, although the term **organism** would seem to imply that.

7. The various functions of plants and animals may be grouped as follows :—

(a) Those essential to the process of waste and repair referred to in §4. The process of waste is mainly one of oxidation and involves the formation of waste-products, partly gaseous, partly fluid, which require to be separated from the tissues: **Respiratory** and **excretory** organs are therefore rendered necessary. The process of repair, on the other hand, not only involves the ingestion of food, but its assimilation, and, therefore, the **nutritive** organs first render the food soluble in such a way that it can be absorbed, and then elaborate it into forms suitable for repairing the waste. **Circulatory** organs are subservient to all three systems.

18

Plants as well as animals are provided with organs of these three categories, but important differences exist as to the nature of the processes discharged by them. The food of plants is drawn from the inorganic world, and consists of salts absorbed in solution from the soil, and of the carbonic acid in the atmosphere. This is decomposed by the combined action of the chlorophyll within the green plant-cells and light—part of the energy of the sun's rays being intercepted by the colouring matter—with the result that carbon is secured for forming simple carbohydrates like glucose, which are afterwards elaborated within the plant-body, while oxygen is disengaged far in excess of that required for the ordinary processes of oxidation.

Thus, from simple compounds, in which little energy is locked up, more complex ones are formed, which are highly energised, and which thus serve for fuel and food to the animal world. Animals are, in fact, dependent either directly or indirectly for their food on the vegetable world, and, as far as we know, not even the simplest among them (apart perhaps from such coloured forms as are referred to in IX, 19) can derive carbon from the inorganic world.

(b) Even more striking differences between plants and animals are to be met with in those organs which relate the individual to its environment; in fact, the organs of locomotion and sensation are generally described as the specially "animal" organs. Not that vegetable protoplasm is destitute of contractility and irritability, but the further elaboration of these essential properties of protoplasm is characteristic of animals, to the exclusion of plants. Nor is it meant that the plant is less adapted to its surroundings on account of the absence of special organs of relation; on the contrary, we shall see that the plant is just as plastic as the animal in its adaptability to its environment. The difference between them in this respect is no doubt to be largely attributed to differences referred to in last paragraph; competition for food on the part of animals having led to the necessity for the complex mechanisms of the muscular and nervous systems.

The functions of the nervous system include all those processes, which, in their elementary forms, appear as the simplest kinds of " reflex action," such as, for example, the contraction of the Amœba on irritation, but pass by easy transitions through higher forms, till those extremely complicated inherited reflexes, which we call "instincts," are reached, and even those higher processes, which we ascribe to the "intelligence" of animals. Such **psychical** processes, from the simplest to the most complex, form the subject-matter of **Animal Psychology** and **Sociology.**

(c) The third group of functions is formed by those which refer to the production of new individuals. Very close parallels are offered by the vegetable and animal kingdoms in respect to these, and multiplication by division (separation of the body into equal parts), the formation of buds (detachment of smaller portions), or, finally, the detachment of reproductive cells, and the formation of new individuals from eggs are to be observed in both. Similar phenomena in connection with this group of functions are also to be studied in both; such as (1) the limited span of life of the individual; (2) the tendency for succeeding generations to increase in number; (3) the tendency of the offspring to inherit the form of the parent, and yet (4) to vary considerably from that form. These phenomena lead us to the discussion of various other aspects of Biology, in the first place, of those which exhibit the relations of present to preceding generations, viz.: Developmental, Palæontological and Taxonomic aspects.

(3) DEVELOPMENTAL AND PALÆONTOLOGICAL.

8. We know that the plants and animals living on the earth at present are the descendants of the immediately preceding gene-

rations. Further, it is evident, on reflection, that few traces of these preceding generations are preserved, such is the destructive power of the minute saprophytic organisms that live by pulling to pieces decaying organic matter. Even the hardest tissues like bone and wood soon crumble to dust, if left to the ordinary process of decay. Now and then, however, the hard parts of a dead animal or plant are preserved, as, for example, when they are washed by a stream to some spot, where they become silted up by the sediment deposited from the stream, and thus protected from the destroying influences to which they would otherwise have been exposed. A reference to the skeletal tissues described for the various groups of animals in the preceding chapter, will make it easy to understand what parts would be likely to be preserved under such circumstances, although under specially favorable circumstances, impressions even of the softest bodied animals are to be found. The material held in suspension by a stream, and deposited as the sediment referred to, is the result of the wearing down or denuding of the land drained by it, and it is obvious that organic remains found in the upper layers of such a deposit must belong to more recent generations than those found in the lower layers.

9. Many of the rocks which form the earth's surface have been formed like the silt referred to, from sediment produced by the ceaseless action of the waves on the sea-coast, or by the denuding action of the various atmospheric influences—rain frost, snow, etc., and swept down by rivers to the shallow waters of the ocean. Such deposits vary much in their character according to the material held in suspension by the water, and result in sandstones, limestones, shales, etc., but they all exhibit a tendency to layering or stratification, depending on the way they were formed, and are, therefore, called **Sedimentary Strata**. The relative age of such strata is determined by their position, and it will be readily understood that the upper-

most or most modern layers will be most like those that we actually see in process of formation, while the oldest or lowest will have been subjected to many changes, such as pressure from those above, upheaval from disturbances of the earth's crust, etc. Now, the examination of the remains of organic life—fossils—preserved in these sedimentary rocks (an important branch of Biology— Palæontology—) shows that only a comparatively thin layer of superficial deposits contains remains of organisms like those at present alive; the deeper down we go, the more do we find different species of plants and animals replacing those familiar to us, until the latter disappear completely and leave us in the presence of an entirely new fauna and flora. The question now arises,—have the species of plants and animals now living on the surface of the globe, and whose remains are found in the superficial deposits, originated entirely independently of the different species found in deeper strata, or are the ancestors of the former really represented among these unfamiliar remains, only unrecognizable at first on account of differences, which the recent forms have accumulated in the course of long ages ? The latter solution of the question is the alternative generally accepted at the present day, and it will be seen that it involves relationship by descent of the present species of plants and animals to some of those of former epochs, others having died out or become extinct without leaving any descendants, just as has been the case within our knowledge of modern species (V, 15, 23 ; VI, 22). Not only does it involve descent, but *descent with modification*, and indeed, when we look at the forms of life in the older rocks we recognize that the modification must have been of a very profound nature. It has, furthermore, been of a definite, orderly character, because we find that the lower classes of plants and animals have appeared before the higher classes—fossil Mammalia, for example, being found only in comparatively recent strata. There has thus been (if we accept the view that the present forms of plants and animals are the modified descendants of other species now extinct) a continuous development from lower to higher forms, presenting in its entirety what is often called the *evolution of organic Nature.* It is, in fact, possible to subdivide the sedimentary rocks according to the prevailing kinds of life during their deposition, into Archæan—which contain few, if any, traces of life—and Palæozoic, Mesozoic, and Kainozoic, in all of which fossils are abundant, but the last of which alone contain any resembling the forms of life at present on the earth. Some information may be gleaned from the accompanying Table as to the characteristic plants and animals of these various periods, and as to the order of their introduction. Not only is the appearance of the life of these periods characteristic, but the classification is assisted, in Europe at least, by the occurrence of conspicuous 'breaks' between them, which indicate, for example, that the Palæozoic strata were elevated into dry land, partly denuded and upheaved, before they again sank; and had the Mesozoic strata laid down upon them *unconformably—i.e.* not in plains parallel to those below as if they had been deposited consecutively. Such breaks do not, of course, occur at the same 'horizon' in other parts of the world, but the classification from the prevailing kinds of life answers all over the world. Each age is again divisible into various epochs, which are marked by more or less distinct breaks in different places, and also by characteristic fossils, so that the geologist is enabled to recognize the rocks of any particular horizon by diagnosing the contained fossils.

10. An important question which comes up in connection with these strata, is that of their age, and, consequently, that of the contained fossils. If we assume that the rate of deposition has been uniform throughout the various strata, then the maximum thickness may be taken as a measure of the relative duration of the time during which they were deposited. The

TABLE OF THE SEDIMENTARY ROCKS,

Showing their relative age and thickness, and their characteristic rock-formations, plants and animals.

AGE OR PERIOD.	EPOCH.		CHARACTERISTIC ROCK-FORMATIONS.	CHARAC. PLANTS.	CHARAC. ANIMALS.
Quaternary or Post-Kain'c.	Pleisto-cene.	Recent Post-glac. Glacial	Indian shell-mounds. Raised beaches, shell-marls,loess Boulder-clay, "till" or drift.	Cultivated plants.	Man.
Tertiary or Kainozoic.	Pliocene Miocene Eocene		Loams, sands, shales, clays, &c.	Angiosperm-ous Phanero-gams.	Mammals.
Secondary or Mesozoic.	Cretaceous.		Lignites, limestones, clays, and sandstones.	Conifers and Cycads.	Reptiles.
	Jurassic.		Marls and limestones.		
	Triassic.		Red sandstones and conglomerates.		
Primary or Palæozoic.	Permian.		Limestones, sandstones and marls.		
	Carboniferous		Coal measures and carboniferous limestones.	Vascular Crypto-gams.	Amphibians and Fishes.
	Devonian.		Sandstones, limestones and shales.	....	....
	Silurian.		Sandstones, limestones and slates.	Algæ.	Inverte-brates.
	Cambrian.				
Archæan and	....		....		....
	Huronian.		Gneiss, mica-slate and various crystalline rocks.		Eozoon (?)
	Laurentian.				

accompanying table shows graphically that the Mesozoic Strata are about one-ninth the thickness of the Palæozoic, and it ought to represent the Kainozoic as about one-fifth of the Mesozoic, and the Post-Kainozoic as one-fifth of the Kainozoic, but the exigencies of printing have made the Kainozoic, and especially the Post-Kainozoic, too thick.

Various calculations have been made as to the absolute time the deposition of the sedimentary rocks has required ; these do not rest on very certain data, but their results come between thirty to sixty millions of years, and are chiefly interesting, as convincing us of the length of time that has elapsed since the appearance of life upon the earth.

(4) Taxonomic or Classificatory.

11. Returning to the question of the descent of present forms of life from past forms, it will be observed that it throws a new light on the terms "allied" and "related," so frequently used in the preceding chapter :—the greater or less resemblance of structure, on which classification is based, is due to community of descent,—to more or less distant blood-relationship.

To facilitate the study of relationship between individuals, *genealogical trees* are framed on documentary evidence, and similar trees have been constructed (going beyond historical times) to show the relationship of nations to each other. The kind of evidence used in the latter case has been partly that obtained from observation of the nations at present, (especially the structure of their language, their folk-lore, etc.,) and partly of a palæontological nature, such as the contents of tombs, and implements, weapons, etc., preserved in the most superficial deposits.

Going further back, however, into geological time, attempts have been made to construct similar trees, to show the relationship of different groups of organic life to each other (**phylogeny**), but it will be understood that although the advance of Morphology is constantly improving our evidence from the living forms, and although hitherto undiscovered fossil species are constantly being unearthed, yet the evidence from the palæontological side must always be incomplete (§ 8). Phylogenetic classifications must, therefore, be tentative, except in the case of such modern groups as are distinguished by the profuseness of their fossil remains. The genealogical tree of the Animal Kingdom might be described as buried, with the exception of the terminal green twigs—the existing species ; to trace the connection between these it is necessary to dig down into the sedimentary strata, and to piece together with infinite patience the fragmentary evidence which we there find. It must be remembered, however, that the principle of correlation depending on the constant association of morphological peculiarities (such as, for example, a ruminant dentition with a particular conform ation of the skeleton of the foot), justifies us in making use of evidence from very fragmentary remains.

12. Such studies may, then, indicate the probable line of de- velopment which has culminated in a particular species ; the discussion of the nature and origin of species will be deferred to a subsequent section. In the meantime, in connection with the foregoing topics the accompanying table may be of use, which gives an approximate estimate of the number of described species in the various classes and orders, and serves to show what groups of animals are in "ascendance" at the present day, as exhibited by their comparative wealth in species, and to call attention to the former wealth as compared with the present poverty of such groups as the Brachiopods, Cephalopods, Crinoids, Pteropods, Trilobites, etc. It makes it apparent, also, how certain highly specialised groups, such as the Teleosts, the Anura, the Passeres, the higher Insecta, preponderate over more primitive allied groups.

TABULAR VIEW OF THE CLASSES AND ORDERS

OF THE ANIMAL KINGDOM,

Showing their relative richness in species.

	LIVING.	FOSSIL.
I.—VERTEBRATA—		
A. Urochorda or Tunicata.		
Class I.—Ascidiacea	270	
Class II.—Thaliacea	30	
B. Cephalochorda.		
a. Acrania	1	
b. Craniota.		
Class I.—Pisces	....	1000+
Sub-class—1. Cyclostomi	17	
2. Elasmobranchii	285	
3. Ganoidei	32	
4. Dipnoi	4	
5. Teleostei		
O. 1. Physostomi	2500	
2. Lophobranchii	125	
3. Plectognathi	177	
4. Anacanthini	370	
5. Pharyngognathi	640	
6. Acanthopteri	30.0	
Class II.—Amphibia.		
O. 1. Urodela	93	
2. Anura	800	
3. Gymnophiona	22	
4. Labyrinthodontia	....	100
Class III.—Reptilia	....	300
O. 1. Chelonia	250	
2. Crocodilia	21	
3. Lacertilia	1250	
4. Ophidia	1000	
Class IV.—Aves	....	300
O. 1. Pygopodes	80	
2. Longipennes	228	
3. Steganopodes	60	
4. Anseres	180	
5. Herodiones	140	
6. Grallae { Paludicolæ / Limicolæ }	470	
7. Cursores	17	
8. Gallinacei	400	
9. Columbae	360	
10. Raptores	540	
11. Macrochires	500	
12. Pici	325	
13. Coccyges	736	
14. Psittaci	400	
15. Passeres	5700	

TABULAR VIEW—(Continued).

	LIVING.	FOSSIL.
Class V.—Mammalia		800
O. 1. Monotremata	3	
2. Marsupialia	149	
3. Edentata	42	
4. Cetacea	155	
5. Sirenia	5	
6. Perissodactyla	23	
7. Artiodactyla	250	
8. Proboscidea	2	
9. Hyracoidea	12	
10. Rodentia	800	
11. Insectivora	150	
12. Pinnipedia	30	
13. Carnivora	340	
14. Chiroptera	415	
15. Lemuroidea	55	
16. Primates	219	
II.—ARTHROPODA—		
Class I.—Crustacea.		
Sub-class—Malacostraca (higher orders)	3525	150
Entomostraca (lower orders)	2075	600
Gigantostraca and Trilobitæ	5	1760
Class II.—Arachnida.		316
O. 1. Arthrogastra (Scorpionina, Phalangina, &c.)	250	″
2. Acarina	900	
3. Araneina	2500	
Class III.—Myriapoda	800	40
Class IV.—Insecta.		2600
O. 1. Thysanura	100	
2. Orthoptera	6000	
3. Neuroptera	1000	
4. Hemiptera	14000	
5. Diptera	18000	
6. Lepidoptera	20000	
7. Hymenoptera	25000	
8. Coleoptera	80000	
III.—VERMES	5500	*
IV.—MOLLUSCA—		
A. Acephala.		
Class I.—Lamellibranchiata	5000	9000
B. Cephalophora.		
Class II.—Scaphopoda	80	160
Class III.—Gastropoda.		
O. 1. Prosobranchiata	9000	6000
2. Pulmonata	6000	600
3. Opisthobranchiata	900	300
4. Heteropoda	60	160
Class IV.—Pteropoda.		
O. 1. Gymnosomata	24	
2. Thecosomata	74	225

TABULAR VIEW—*(Continued)*.

	LIVING.	FOSSIL.
Class V.—*Cephalopoda*.		
O. 1. Tetrabranchiata	4	4200
2. Dibranchiata	140	310
V.—MOLLUSCOIDEA—		
Class I.—Brachiopoda.		
O. 1. Testicardines	80	2200
2. Ecardines	30	400
Class II.—Polyzoa.		
O. 1. Phylactolæmata	50	
2. Gymnolæmata ....,......	690	1850
VI.—ECHINODERMATA—		
Class I.—Holothuroidea......................	440	
"　II.—*Echinoidea*	300	2000
"　III.—*Ophiuroidea*........................	700	
"　IV.—*Asteroidea*...........................	500	
"　V.—*Crino.dea*	430	1170
"　VI.—*Cystoidea*		140
"　VII.- *Blastoidea*		90
VII.—CŒLENTERATA—		
Class I.—Ctenophora :........................	45	
"　II.—*Hydrozoa*	1100	
"　III.—*Actinozoa*	1800	1800
VIII.—PORIFERA	600	800
IX.—PROTOZOA—		
Class I.—Sarcodina.		
O. 1. Rhizopoda	700	1500
2. Heliozoa................	40	
3. Radiolaria	2600	400
Class II.—Sporozoa..................	55	
"　III.—*Mastigophora*	290	
"　IV.—*Infusoria*	500	

* The fossil remains of Vermes, which consist chiefly of the tubes of tubicolous Annelids, and the jaws of Errantia, are too uncertain in character to allow of numerical estimate.

13. The preceding paragraphs have been devoted to the relations existing between different generations of organisms; no less important are the relations of the individual or the species to other organisms and to its inanimate surroundings. Questions of this character have been styled **Mesological,** as dealing relation to the "environment," but it will be more convenient to deal separately with the relations to the different elements of the environment, discussing, in the first place, the distribution of animals in space, which we shall find to be explained partly by the past configuration of. land and water, and partly by climatic influences. These aspects of Biology are, therefore:

(5) Geographical and Climatic.

We are apt to think of the present distribution of land and water as something unalterable, and of climate as chiefly a matter of latitude and longitude, but a little consideration will show that not only is the latter dependent on the former, but that both have been, again and again in the history of the earth, subjected to change. In respect to the one point, we have merely to compare the mild winters of Britain with those of Labrador, or our own with those of the Riviera. The differences are attributable, in the one case, to ocean currents, in the other to our situation in the heart of a great continent. In respect to the other point we can find evidence of profound changes at our own doors. Wherever we find sedimentary rocks there we realize that the land in question has been submerged in the ocean, and often that this has occurred, not only once, but repeatedly in the course of geological time. Ontario, for example, is everywhere underlaid by sedimentary rocks; those which contain fossils were evidently formed at the bottom of Cambrian or Silurian seas, but the fact that we do not find any trace of the Upper Palæozoic or Mesozoic or Kainozoic strata covering them indicates that it has been long exposed to the denuding effects of the atmosphere. Again, the superficial deposits which cover these old formations,

and which belong to Post-Kainozoic Age, show that compara·
tively recently, Ontario, with the greater part of North America
as far south as Washington, was covered by an Ice-Sheet, so
that then the whole country resembled the present desolate and
lifeless interior of Greenland. Can it be doubted that such
changes taking place all over the world must have had the most
important influence on the distribution of life?

While we recognize a distinction between tropical, temperate
and arctic faunas and floras, and are obliged to infer climatic
changes if we find that Magnolias flourished in Spitzbergen during the Miocene epoch, and that Hippopotamuses wallowed in
British rivers in the intervals of successive Ice-Sheets, yet we
must be careful not to attribute too great an influence to climate
alone, without taking into account the geographical changes to
which the climatic ones are secondary.

14. That such changes have been both recent and profound
will now appear from some of the facts of distribution already
cited, and the explanation of these offered by palæontology.
First, however, let us look into the causes and effects of that
Glacial Period, during and after which the surface deposits of
Ontario were formed.

There are certain regularly recurring astronomical conditions,
which affect the relative length of winter and summer, and
which are dependent (1) on the relative proximity of the poles
to the sun during these seasons, and, (2) on the amount of
eccentricity of the earth's orbit. The relative proximity to the
sun in winter of the North and South Hemispheres is reversed
every 10,500 years, but the amount of eccentricity varies much
more slowly. Now, it is at its minimum; between 100 and
200,000 years ago it was very great, with the result of alternating periods of 10,000 years of long, cold winters, and short,
hot summers, with the reverse. It is believed that these con·
ditions, occurring simultaneously with large elevations of land

within the Arctic Circle (where, as we know from fossils found, there had been warm seas during the Miocene epoch,) had the effect of producing the glaciation. It must not be supposed that the Glacial Period lasted with undiminished severity during all this time; we have evidence of interglacial milder periods, when the Ice-Sheet retired, and the life, which had been driven towards the south, again crept northward, but these very alternations must have been a fruitful source of change in the fauna and flora of the countries subjected to them.

15. Apart from the combined effects of the astronomical conditions referred to, and the raising of the land towards the North Pole, let us consider the effects of a general elevation of the sea-bottom outside the present shore-line, to an extent far less than we know to have taken place within comparatively recent times.

The summits of some of the Alps—the Dent du Midi, for example, 10,770 feet—are formed of Nummulitic limestone, a formation consisting largely of shells of Foraminifera, deposited during the middle Eocene epoch, and elevated since then. Now, we know from hydrographic researches that if an elevation of the shallower waters of the world to half this extent were to take place, that it would entirely alter the configuration of the land. Not only would Great Britain be continuous with France, Denmark and Norway, but the whole of Europe, Africa and Asia would form one continuous Continent. It would be even possible to travel by land from England to Canada, viâ Iceland, Greenland and Labrador, and then on to Siberia by Alaska. The only large tracts of land remaining unconnected with this vast Continent would be Australia, which would still be an Island, and Madagascar, which would still be separated from Africa.

Ample evidence exists, and some of it has been already referred to (VI. 30), which shows that changes of such a character have occurred, not necessarily simultaneously, but at least at different times. The Camel family, e.g., flourished in North

America in Miocene times and spread thence to South America and to Asia where they had not previously existed, but to which the distribution of the modern form is limited. Again, the primitive forms of the Tapir family lived in Europe during the early part of the Eocene epoch, they spread during the later Eocene and Miocene epochs to North America, where, however, they died out without leaving any descendants, but they persisted in Europe, where they gave rise to the modern Tapirs, and these again spread to Asia and South America, where alone the modern species occur.

That Madagascar was at one time connected with Africa, we gather from the survival on the mainland of some of those peuliar lemurine forms, whi ch are so characteristic of the Island. But these have almost been exterminated from Africa by the contact of the higher Eutheria, so that we conclude that the connection with Africa was submerged before the influx of the higher mammals from Eurasia. The persistence of the Prototheria and Metatheria in Australia is similarly to be attributed to its long independence from Asia.

The fauna of Great Britain, finally, can only be accounted for on the assumption that since the Glacial period it has been entirely submerged, raised again, and populated from the mainland, but again submerged before this repopulation could become complete. The freedom of Ireland from reptiles is not to be attributed to any unsuitability of climate, but to the fact that the channel was not dry land sufficiently long to allow these forms, which are slow at distributing themselves, to reach it.

16. Such facts lead us to the position that the geography of the past has had the most important influence on determining the geographical zoology of the present. Each species has a certain area of distribution, which is called its habitat: it tends to widen this area, as far as the conditions of existence will per-

mit, but the most effectual barriers are high mountain-chains and deep seas. By mapping out the ranges of different species it is possible to mark out certain zoogeographical regions in which the simultaneous occurrence of the same or nearly allied species confers a certain similarity of appearance or facies to the fauna. A glance at these regions, as laid down by Dr. Wallace in the accompanying map, will shew that they agree pretty well with our ordinary geographical divisions, except that great natural barriers cut off the Ethiopian region from North Africa, the Oriental from the Palæarctic, and the Australian from the Oriental ; these are the Sahara, the Himalaya, and the very deep sea along " Wallace's line," between Bali and Lombok, Borneo and Celebes. This is certainly a most instructive illustration of the effect of a deep-sea barrier, and of the fact that mere climate does not determine geographical distribution. From Bali to Lombok is hardly twenty miles, and yet there is more difference as to the fauna between these two islands, the climate of which is practically identical, than there is between Canada and Florida.

17. Although the distribution of aquatic animals admits also of geographical treatment, it is obvious that the barriers to their spreading far and wide must be of very different character, and, apart from those forms confined to the shore by limited loco-motive powers, must be largely due to ocean currents, tempera-ture and other surrounding conditions. Other elements of interest, however, exist in connection with them—their occur-rence in fresh or in salt water, and their bathymetrical distri-bution—i.e., the depth at which they occur.

The greatest profusion of life is found in the shallow waters of the shore-zone, but recent observations have shown that it also abounds in the depths of the sea, even down to four or five miles. Again, there are so-called pelagic forms which live on the surface-waters of the open sea, and are only occasionally
19

driven upon the shore. It is believed that the littoral fauna originated from the pelagic in the first place, while many pelagic forms, such as the whales, are undoubtedly derived secondarily from shore forms. The origin of fresh-water faunas is another point of great interest; certain stages through which they must have passed may be seen in the mouths of rivers where different species extend upwards to a greater or less distance, according as they are more or less tolerant of the diminished salinity of the water. That part of the fresh-water animals are so derived may be gathered from such cases where they exhibit little or no difference in their new habitat, (VI, 34, VII, 13). There are many curious facts in the distribution of the fresh-water fishes, which indicate the early separation of some groups from marine forms, and which can only be explained by realizing that changes in the configuration of the land must have involved many changes in the great water-courses.

In inland lakes we have also littoral, deep, and pelagic faunas, the study of the nature and origin of which offers a wide field of inquiry in Ontario. Before leaving the subject of the distribution of animals in water, it may be noted that not only do ocean currents affect the ranges of aquatic, but also of terrestrial organisms, so that, apart from their climatic influence on these, they may frequently transport animals and plants to new points at considerable distance from their original homes.

---

(6) Relations of Animals to the Conditions of Existence and to each other

By climate we mean temperature, pressure and other atmospheric conditions, which do not seem to determine any conspicuous differences of form, but there are other conditions of existence which do seem to be always associated with corresponding differences of structure ; such are the association of underground life with the absence of eyes and of colour (II, 79 ; III, 12) ; and that of the shape of the body with the medium and mode of locomotion (Fig. 194). These offer a very tempt-

Fig. 194.—Harp Seal. Phoca grœnlandica'(from Brehm).

ing field for investigation, and cause us to enquire how changes in surrounding conditions and in habits affect animal life.

19. Reference has already been made to the fact that organisms tend to increase in number in geometrical progression. Sometimes countless eggs are produced by a single individual, but in such cases few of these reach maturity, and it is only now and then that we are startled by the disproportionate development of some one species, disturbing the course of Nature. When, however, one generation exceeds the previous one, even in the proportion of two to one, it is evident that there will soon be a competition or struggle between the individuals for suitable food, and that the strongest and best adapted to survive will be those, on the whole, that do survive. Along with this we have to take into account the tendency of organisms to vary and to transmit their peculiarities to their descendants (§ 7). Those individuals whose organs vary in such a way as to adapt them better, however little, to their special circumstances of life, are,

therefore, rendered better fitted to survive, and will not only do so, but will transmit their peculiar variations to their off-spring ; selection being thus made by Nature of the individuals best fitted to the existing conditions of life.

20. If those conditions were stable, a persistent equilibrium of organic nature would be established, and no opportunities for the development of new species would arise, but the conditions of life are never stable, and consequently we have a greater or less degree of instability of the balance of life, depending on the greater or less changes taking place in the conditions of exist-ence. It is obvious that any apparent equilibrium would be upset by changes, however gradual, of the character indicated in § 14 and 15, and, therefore, that such changes must be regarded as those most operative in the production of new forms. A species subjected to such changes must either produce variations fitted to cope with them, or else become extinct, perishing at once, or struggling on under the unfavourable influences, till a depau-perate condition, such as has been observed in many fossils, as well as in living forms, is gradually reached. If it is a form capable of adapting itself, by assuming new habits, it is possible that (as we see in varieties spread over a wide area) this accom-modation should proceed along different lines with different varieties, and therefore, that several new forms would result from the original species. It is evident that the intermediate forms be-ing liker the original in character would tend to be eliminated by their relative unsuitability to the new conditions, and thus, instead of a series of varieties, we should have two or more new 'species' distinct from each other. Another circumstance, tending in this direction, is that individuals of a variety tend to keep together, with the result that comparatively near species become incapable of crossing, so as to form a mixed race. This was at one time considered to be an essential feature in the definition of a species, but several examples are now known of undoubtedly distinct species, which do form such crosses, where their areas of distribution come into contact.

21. The theory stated in the preceding paragraphs is that of the Origin of Species by Natural Selection, associated with the names of Darwin and Wallace ; it will be observed that while resting upon the large amount of variation offered, it does not attempt to explain the cause of such variation. This is attributed by certain American zoologists,—of whom Cope is the chief representative—to the direct action of the environment, for example, the gradual preponderance assumed by the central digits in the Ungulates would be explained by the greater strain received by those reaching the ground. Strict Darwinists do not consider such an explanation to be sufficient, because there are many instances of protective resemblance and mimicry where just as remarkable modifications of form are to be met with, which could not be attributed to such a direct action of the environment. On the other hand, we have met in the preceding chapters with so many instances of the adaptation of the organism to its habits (*vide* Index, adaptation) that it seems difficult to believe that such remarkable correspondence should only be the result of selection from variations tending to occur in every direction.

22. Not only have we to contemplate the effect of competition between different individuals as favourable to change in organic nature, but also the competition of different species. This is readily seen in a garden which, if left to itself, soon becomes overgrown with weeds—those forms of plants which have either better means of protecting or dispersing themselves, or are able to cope with less favourable conditions of life than the plants formerly cultivated. It is easy to see, that such competition, where species come into contact for the first time, may lead to extinction of the form less able to protect itself; indeed the extinction of past forms of life is to be explained partly by such competition, and partly by reason of some forms being specialised to such a narrow range of conditions, that they have become

incapable of adapting themselves to any changes. On the other hand, we may explain to ourselves the persistence of certain types (VIII, 25), either by the little specialised character of their requirements, or by their gaining superior means of protection, as for example, the adoption of a subterranean life, by terrestrial animals, or their retirement to the recesses of deep forests.

The competition of different species is interesting in another aspect—the regulation of the balance of life. It is obvious that the number of carnivorous animals is regulated by the number of phytophagous forms on which they feed, and that these, again, are dependent on their food-plants.

23. While reference is being made to species which live in competition with each other, it must be recalled that many instances of association for mutual advantage are to be found in Nature. One of the most striking of these is afforded by some Sea-anemonies, which fasten themselves over the abdominal region of certain Paguridæ. They serve in place of the sheltering shells, which most of the genera select (VII., 12), and, in addition, protect the crab by reason of their thread-cells; in return, they are furnished with locomotive facilities not usually enjoyed by their relations. Other examples of such mutualism, or symbiosis, are not uncommon.

24. Still more common, however, are the cases of partial or complete parasitism, which are to be met with in all the sub-kingdoms. (See index). The various grades of parasitism offer such easy transitions from completely normal to much reduced organs, that we are tempted to seek an explanation for these in the direct action of the environment. It is especially the locomotive and digestive organs which exhibit such reduction, and the disuse of these seems to offer a rational explanation of their condition.

25. In the course of disappearance of organs which are under-

going reduction, and before all trace of them disappears, they are styled "rudimentary" organs; such are the rudimentary metapodials of the horse. Organs like these receive their only satisfactory interpretation when we look at them in the light of the doctrine of descent with modification.

26. The active relations between different species discussed above, lead us to the consideration of the remarkable phenomena of protective resemblance and mimicry (VII., 26, 33)—phenomena which appear to be explainable only by natural selection. The word mimicry conveys a striving after similarity, which is entirely at variance with such an explanation, and the term is now really taken to mean the preservation by Nature of variations in the direction of resemblance to some other animal, protected either by offensive or defensive weapons. Cases of such protection are recorded above. A further illustration is afforded by certain South American butterflies belonging to the family Pieridæ, which "mimic" those of another family, the Heliconidæ, protected from insectivorous birds by their offensive odour and taste.

Protection may also be secured, however, by resemblance in colour, or form, or both, to surrounding plants or inanimate objects. The winter white coat of Arctic animals, the colours of the nests and eggs of birds, the form of the leaf- and walking-stick insects (VII, 23), are all to be explained in the same way. Occasionally such likenesses are employed, not for defensive, but for offensive purposes. One species of the predatory genus Mantis (a member of a family allied to the Phasmidæ) resembles an orchid which is visited for its honey by bees, and a species of spider has been observed, which, in the attitude in which it waits for its prey, has the innocent appearance in colour and form of a bird-dropping!

Colouration, in fact, is very generally protective in its function, and the transparency of most pelagic animals, the changeable hues of the chamæleon and the tree-toad (which are effected reflexly through the nervous system), are to be explained on the same principles. This does not exhaust its functions ; it would appear with gregarious animals to be useful in recognition of other individuals of the same species, also for warning, as in the case of the brilliantly coloured coral-snakes (*Elaps* IV., 21). That such warning colours are not unfavourable to their possessors, might lead us to suppose that the rattlesnake's rattle (IV., 22) is a similar warning to enemies not to interfere.

(7). RELATION OF ANIMALS TO PLANTS.

26. Some of the phenomena recounted above serve to recall that plants are not destitute of defences provided by natural selection in relation to phytophagous animals. Such are the spines, and the bitter, acrid and poisonous excretions, such as tannin, which the best protected possess. It is only in comparatively rare instances that plants can be said to assume the agressive, and seize upon opportunities for securing food by absorption of proteids from animals, instead of manufacturing them with the aid of the nitrates of the soil. Such cases do occur in the **carnivorous** plants, like the Venus' fly-trap (*Dionœa*), and the more familiar pitcher-plants (*Sarracenia*) and sun-dew (*Drosera*). It must not be forgotten that there is a whole division of the vegetable kingdom—the Fungi—which, being destitute of chlorophyll, require to get their nourishment from previously formed organic matter. They are either saprophytic or parasitic forms, the latter sometimes so actively aggressive, as to be one of the most fruitful causes of disease and death, both in the animal and vegetable kingdoms.

27. But there are also instances of plants and animals deriving mutual advantage from living together, which recall the cases of symbiosis of animals. Such are afforded by the "yellow

cells" of the Radiolaria, which seem to prosper in their hosts and to occasion them no inconvenience by their presence. Possibly they furnish oxygen to the tissues, and utilize themselves the waste products, including carbonic acid, excreted by the animal tissues.

28. The most obvious instances of the association of animals and plants for mutual advantage are to be met with in the reproductive phenomena of the vegetable kingdom, chiefly in those connected with the fertilisation and distribution of flowering plants. It is now known that by far the greater number of Phanerogams with conspicuous flowers are fertilized by ins ct-agency, and that the secretion of nectar, the colour and often the form of the flower are so many inducements, acquired by natural selection, to insects to visit them, while, on the other hand, many peculiarities of form, or of the relative development in time, of parts of the flower are so many obstacles to self-fertilisation. So close is the relationship between flowering plants and the insect-world that they may be said to have been correlatively developed.

Similar mutual advantages exist in the relation of frugivorous birds to fruit-bearing plants; the distribution of such plants has been shown to be largely effected by the birds in question, and the bright colours of fruits as well as their sweetness receive a partial explanation in this way. Other examples of the co-operation of animals in the distribution of plants are furnished by those whose fruits (burrs), being clothed with hooks or spines, adhere tenaciously to the coats of various mammals, and thus secure a wider range.

29. Such considerations as those above emphasize the interdependence of the vegetable and animal kingdoms, and the necessity of studying the phenomena of both in connection with each other. This is rendered more imperative when we come to look at the economical aspects of Biology, and realize that the

diseases of cultivated plants and animals are frequently avertible by a proper knowledge of the life-history of the organisms in question. Many of these diseases are merely phases of the struggle for existence, which occurs not only between individuals of the same species but between species of the most different character. Other diseases are due to unfavorable surrounding conditions, which a due attention to physiology may make it posssible to remedy. Such disturbances of the normal or healthy processes of life form the subject-matter of Vegetable and Animal **Pathology.**

For further information on these as well as on the other topics discussed in this chapter, the student must, however, consult books which deal with them specially. The object of the present chapter has chiefly been to show the relations of facts scattered throughout the preceding chapters, and to indicate as far as possible the general principles which have been reached by zoologists.

INDEX.

296

20

314

NOTE.

The following suggestions as to the order and scope of the practical lessons which should precede the study of the Text-Book as a whole may be of use to Teachers. The References are to Chapters and Sections of the Text-Book.

1. Thorough examination of the external form (I. 1-5), the gills (56) and the viscera (49-65) of some common fish.

2. Study of the prepared skeleton—preferably of the catfish, on account of the assistance to be obtained from the figures (I. 10-30).

3. Demonstration of the arrangement of the muscular and nervous systems (31-35, 37) and the sense-organs (41-3) as far as these can be studied *without* the aid of the microscope.

4. Comparison of the structure of the frog (III. 16-22) with that of the fish, and with that of the Menobranch described in III. 1-10. The skeleton of the pectoral and pelvic girdles and of the appendages of the frog, should be compared with the figures of the same parts in the Menobranch (III. 5).

5. Examination of the external form of a Turtle (IV. 1-12) and a Snake (17-20).

6. Examination of the structure of a pigeon or a fowl (V. 1-3, 5-12).

7. Study of the skeleton (VI. 5-7), also of the teeth and viscera (11-16) of a cat or dog.

8. Study of a crayfish as a type of the Arthropods (VII. 1-11).

9. Comparison of the Crayfish with an Insect—Grasshopper, Cricket, or Cockroach (19-26)—also with a Millipede (23) and a Spider (17).

10. Examination of an Earthworm and Leech (VIII. 1-6).

11. Study of a fresh-water mussel and s pond-snail (VIII. 14-20).

12. The principles of Zoological nomenclature (II. 1-8) as illustrated by some of the common fresh-water fish, such as the sucker and herring (9-11), bass and perch (17-18).

13. Study of an Amœba or Paramœcium as a type of an unicellular animal.

14. Cells and tissues of the higher animals (I. 6-9) and the microscopic structure of their organs (I. 11, 31, 38, 44, 48, II. 2, VI. 4).

15. Development of the eggs of fish in the Spring (I. 66), and the metamorphosis of the tadpole (III. 21).

In addition to practical lessons such as those indicated above, the following topics ought to be illustrated by figures, diagrams, etc :—

The modifications of the form of the body in Vertebrates in connection with different methods of locomotion—Figs. 30, 39, 41, 43, 65-68, 71-2, 74-5, 78-9, 81, 84-90, 92, 95, 98-9, 106, 108, 111-3, 115, 116-7.

The characters of the orders of Mammalia (VI. 17-41) [large type] and their geographical distribution—compare X. 16-16 and map.

The succession of the fossiliferous rocks (p. 271) and the importance of the Ganoids (II, 24), the fossil Reptiles (IV. 22-26), the Trilobites (VIII. 15), Brachiopods (VIII. 25) and Corals (IX. 6), from a Palæontological standpoint.

www.ingramcontent.com/pod-product-compliance
Lightning Source LLC
Chambersburg PA
CBHW021503210326
41599CB00012B/1117